Wittgenstein

Also by Pasquale Frascolla

WITTGENSTEIN'S PHILOSOPHY OF MATHEMATICS (1994)

UNDERSTANDING WITTGENSTEIN'S TRACTATUS (2006)

Also by Diego Marconi

LEXICAL COMPETENCE (1997)

WITTGENSTEIN (*editor,* 1997)

LA PHILOSOPHIE DU LANGAGE (1997)

FILOSOFIA E SCIENZA COGNITIVA (2001)

PER LA VERITÀ (2008)

Also by Alberto Voltolini

GUIDA ALLA LETTURA DELLE RICERCHE FILOSOFICHE DI WITTGENSTEIN (1998)

HOW FICTA FOLLOW FICTION (2006)

I PROBLEMI DELL'INTENZIONALITÀ (*with C. Calabi,* 2009)

Wittgenstein

Mind, Meaning and Metaphilosophy

Edited By

Pasquale Frascolla
University of Basilicata, Italy

Diego Marconi
University of Turin, Italy

and

Alberto Voltolini
University of Turin, Italy

palgrave
macmillan

First published 2010 by
PALGRAVE MACMILLAN

Palgrave Macmillan in the UK is an imprint of Macmillan Publishers Limited, registered in England, company number 785998, of Houndmills, Basingstoke, Hampshire RG21 6XS.

Palgrave Macmillan in the US is a division of St Martin's Press LLC, 175 Fifth Avenue, New York, NY 10010.

Palgrave Macmillan is the global academic imprint of the above companies and has companies and representatives throughout the world.

Palgrave® and Macmillan® are registered trademarks in the United States, the United Kingdom, Europe and other countries

ISBN: 978–0–230–21941–0 hardback

This book is printed on paper suitable for recycling and made from fully managed and sustained forest sources. Logging, pulping and manufacturing processes are expected to conform to the environmental regulations of the country of origin.

A catalogue record for this book is available from the British Library.

A catalog record for this book is available from the Library of Congress.

10 9 8 7 6 5 4 3 2 1
19 18 17 16 15 14 13 12 11 10

Printed and bound in Great Britain by
CPI Antony Rowe, Chippenham and Eastbourne

Contents

Part III Metaphilosophy

Contributors

William Child is Lecturer in Philosophy at University College, Oxford, UK

Pasquale Frascolla is Professor of Philosophy of Language at the University of Basilicata, Italy

Hans-Johann Glock is Chair in Philosophy at University of Zurich, and is currently President, Philosophical Society Zurich Switzerland

Paul Horwich is Professor of Philosophy, New York University, USA

Sir Anthony Kenny (retired) University of Oxford, UK

Diego Marconi teaches Philosophy of Language at the University of Torino, Italy

Eva Picardi is Professor of Philosophy of Language, University of Bologna, Italy. She is also Director of the Doctoral Programme "Mind, Logic, and Language"

Joachim Schulte teaches at University of Zurich, Switzerland

Frederick Stoutland is Permanent Visiting Professor, University of Uppsala, Sweden

Alberto Voltolini is a Philosopher of Language and Mind, and is Associate Professor at the University of Turin, Italy

Meredith Williams is Professor of Philosophy at Johns Hopkins University, USA

Timothy Williamson is Wykeham Professor of Logic, New College, Oxford, UK

Introduction

Pasquale Frascolla, Diego Marconi and Alberto Voltolini

There is little doubt that interest in Wittgenstein's philosophy has been declining in the last 20 years. This is particularly true with analytic philosophers, who had regarded him as one of their founding fathers together with Frege and Russell. However, while Frege's writings are painstakingly investigated and Russell's paper *On Denoting* is commented upon line by line, nowadays few read *Tractatus logico-philosophicus*, although the voice it represents in the early twentieth-century debate on the nature of language, logic, and mathematics could hardly be disregarded. As to the 'later' Wittgenstein, the author of *Philosophical Investigations*, for most of today's young American philosophers he represents a paradigm of bad philosophy, to be confined to the despicable company to which Dewey and Heidegger also belong.

In recent times, the rejection of Wittgenstein by the analytic mainstream may have been accentuated – *a contrario* – by the relative success of so called 'New Wittgensteinians,' the interpretive trend represented by Cora Diamond, Jim Conant and others, who tend to overemphasize the opposition between Wittgenstein and the dominant style of analytic philosophy (beside putting forward rather bizarre exegetical claims). Thus Wittgenstein comes out as the monopoly of an embattled, yet insular minority, which provides everybody else with an additional reason for forming a bad opinion of him.

One must admit, however, that such an opinion had been around before. More and more often, Wittgenstein had been seen as needlessly cryptic, rambling, neurotically incapable of philosophical professionality (of which inability he would have been proud, one must say). The views he had defended, to the (limited) extent that they were judged intelligible, were mostly rejected: there is no categorical distinction between reasons and causes, indeed, reasons *are* causes (Davidson);

the domain of the mental is only indirectly connected to public behavior, such a connection having no conceptual import; a private language is perfectly possible, and the arguments by which Wittgenstein rejects it should be thrown into the crowded garbage can of verificationism (Fodor). A whole discipline – cognitive science – was born by choosing as its field of investigation a domain of entities about which Wittgenstein believed that almost nothing could meaningfully be said. And one should not forget how insistently it is repeated that philosophy is not specially about language (though it is not always specified what it *is* about, instead).

Sometimes, one has the suspicion that the two parties – the Wittgensteinian and the anti-Wittgensteinian – are simply talking past each other. Did Wittgenstein reject a private language in the very same sense in which it is countenanced by Fodor? Or – to point towards an altogether different subject – does Quine's rejection of the analytic/synthetic distinction *eo ipso* dispose of Wittgenstein's distinction between grammatical and empirical propositions? Even when no misunderstanding is in question, is the rejection of Wittgensteinian views – such as the distinction between causes and reasons, or the idea of an intimate link between a mental state and its manifestation – ultimately grounded in convincing arguments, or is it merely a matter of philosophical fashion?

Aside from such theoretical disagreements, important as they are, the main reason for the present rejection of Wittgenstein's thought may hinge on the sociology of academia no less than on philosophy. Wittgenstein consistently claimed that philosophy is an activity aiming at clarifying our thoughts, dismantling the traps into which language makes us fall all the time, and (consequently) dissolving philosophical problems. Although there is a conspicuous debate on whether he himself suffered from a form of self-deception on this concern, he indisputably maintained that in philosophy there are no theories: the urge to produce philosophical theories is born of a misunderstanding of the relation between philosophy and science, and whatever theories are produced are constantly wrong. By contrast, today's analytic philosophy is intensely theory-oriented: Kripke's *theory* of reference and Rawls's *theory* of justice are paradigms of philosophy. Every young philosopher dreams of becoming a celebrity by solving one of the canonical philosophical problems by way of a brilliant theory.

Wittgenstein opposed the spirit of his philosophy to that characterizing 'the great stream of European and American civilization to which we all belong', which 'issues in the construction of ever wider and ever more complicated structures'. There is little doubt that, in today's Anglo-American philosophy, the construction of structures takes center stage: ever more complicated, though perhaps not always wider structures. To the extent that it is reflected in academia, 'the great stream of European and American civilization' focuses on the scientific paper, specialized journals, referees, competition among theories and among the personalities with which they are identified. Whatever their strictly cultural merits, such practices facilitate job and resource allocation, they confer identity and make philosophy inherit part of the prestige that goes with the hard sciences. That is why this is the dominant model, even for philosophers. Now, for all such things Wittgenstein would have felt deep horror. Was he right, wrong? Could it be that he was partly right and partly wrong? Can a sober evaluation be applied to the claims Wittgenstein apparently put forth, by investigating more neutrally than it was done in the past whether at least some of them still resist philosophical criticism?

With these ideas in mind, in Spring 2006 we organized in Reggio Emilia (Italy) a conference on 'Is There Anything Wrong with Wittgenstein?' We invited to the conference some eminent commentators of Wittgenstein's philosophy as well as some contemporary philosophers who, independently of whether they can be regarded as 'Wittgensteinians', have at least a benign attitude with respect to Wittgenstein's thought, in order to discuss to what extent that philosophy can still be considered to be alive. This volume can be regarded as a natural development of that conference. We don't know whether the above questions have found a definite answer. Yet we hope that at least the way to find an answer has been traced. In what follows, we have organized the chapters that constitute this volume in three different sections, which in our opinion represent the main subjects on which Wittgenstein's reflections have contributed to philosophical enlightenment: mind, meaning, and metaphilosophy. Let us briefly summarize the chapters' content, according to the different sections in which they are located.

The section on mind opens with a chapter that addresses one of Wittgenstein's convictions that are nowadays hardest to swallow, namely his idea that mental states are intimately linked to behavior. William Child shows that for Wittgenstein we undoubtedly apply a mental notion even in cases in which there are no behavioral manifestations of the corresponding mental state, although it is also clear that for him such an application depends on applications in which the corresponding mental states are openly manifested. Child wonders which of three models best accounts for such a dependence: a nonverificationist model according to which the 'unusual' application of the mental notion in question is as legitimate as the 'usual' one, an antirealist model, according to which the former application stretches the notion's boundary, and a verificationist model, according to which the application is based on finding a physical trace – typically, a brainy trace. Child concludes that Wittgenstein may be seen as essentially wavering between the first two models.

Joachim Schulte addresses the famous sections of the *Investigations* devoted to the clarification of what reading is. By stressing how those sections constitute a unitary block substantially derived from a previous reflection in the *Brown Book*, Schulte shows that Wittgenstein wants to focus on human reading as a mental activity which stands in between a mechanical way of reading that machines actually perform and a para-mechanical (mis)conception of reading which interprets reading as essentially consisting in a psychic experience vivifying dead symbols. One might gloss that human reading, as an operation of deriving sounds from written symbols in conformity with some rule of derivation, is a paradigmatic case of obeying a rule blindly. Yet, Schulte notes how for Wittgenstein reading also involves some Gestalt-like operation by which the written symbols are seen as having a certain unity and familiarity. According to Schulte, although Wittgenstein was tempted to regard that operation as consisting in having a *sui generis* experience of some kind, he resisted the temptation: having a determinate (*bestimmt*) experience while reading does not amount to having an experience belonging to a certain sort.

As known, the so-called 'standard view' of action, according to which, roughly, actions are to be conceived of as bodily movements caused and rationalized by the agent's beliefs and desires, understood as inner states or events in her mind/brain, gradually replaced the Wittgensteinian orthodoxy based on the principle that reasons cannot be causes of actions. One of the aims Frederick Stoutland pursues in his chapter is that of throwing light on the true role that Davidson's work played in that epochal transition. As Stoutland notes, Davidson himself stressed how fundamental, in the conflict between his conception and the Wittgensteinian viewpoint, the issue was of whether rational explanations of human behavior are causal explanations. As a result of his deep investigations, Stoutland makes a distinction between two interpretations of Davidson's account of the rational explanation of action: one that is usually taken as a source of the standard view, and another – the right one in Stoutland's opinion – on which the distance between Davidson's conception and the best Wittgensteinian accounts is definitely smaller than it is usually taken to be. One of the most important points of convergence which clearly emerges from Stoutland's analysis is that, contrary to physicalists and dualists, both Davidson and the Wittgensteinians construe beliefs and desires not as states either of the brain or of the mind, but as states of whole persons.

Alberto Voltolini ends the 'mind'-Section by stressing two points. First, the 'later' Wittgenstein's appeal to the private language argument against a pre-linguistic conception of thought is vitiated by an ungrounded assumption. *Second,* the famous criticism the 'later' Wittgenstein apparently addressed against a younger self about a conception of mental states as endowed with both original and intrinsic intentionality is also similarly vitiated. In the former case, the assumption is the normativity of language. In the latter case, the assumption is the factual nature of mental states prompted by their representational character. Once those assumptions, especially the latter one, are dropped, there is room not only for a pre-linguistic conception of thoughts, but also for thoughts endowed with both original and intrinsic intentionality. Such a theoretical space, Voltolini concludes, is not utterly anti-Wittgensteinian. For the mature conception one can draw from Wittgenstein concerning the relationship between (public) language and thought is that complex thoughts, but not simple thoughts, need (public) language to be articulated.

According to a familiar picture, Wittgenstein insisted on both the rule governed nature of language and the idea that linguistic rules are conventions, that is, rules that are both arbitrary and intersubjectively shared. Against both Chomskyan and (more profusely) Davidsonian semantic individualism, Hanjo Glock intends to vindicate the Wittgensteinian picture: meaningful language requires the existence of conventions. Particularly, language requires *lexical* conventions, that is, 'norms which fix the meaning of particular expressions'. Without such norms there would be no correctness conditions for the use of words, hence no meanings. Glock first defines a 'weak' notion of convention that is intended to overcome the objection that for conventions to be possible, language must already be in place (for conventions are said to require that the participants have beliefs and attitudes towards the beliefs and attitudes of others, and such attitudes, in turn, require language). He then goes on to examine (and reject) the individualists' arguments. His general attitude towards the individualistic picture can perhaps be summarized by his concluding remarks: the individualistic picture 'puts communication at the mercy of coincidence. It also puts it at the perhaps even more tender mercy of human intelligence. It overintellectualizes linguistic communication by demanding speakers to engage constantly in highly complex interpretations. The communitarian picture is far more realistic.'

As is well known, Paul Horwich supports a use theory of meaning. In his article, he reformulates the theory and defends it against criticism implicit in Kripke's 'Wittgensteinian' scepticism and general views about meaning, rules, and rule following. In the Appendix, Horwich attempts to show that his views are consonant with Wittgenstein's; indeed, more so than some of the views that Kripke attributes him. In some cases, he relies on the standard picture of Wittgenstein's thought: thus he states his agreement with Wittgenstein on deflationism about truth (quoting PI §136 in support), on the idea that meaning is use, on the variety of word use, and on rules being sometimes followed implicitly, that is, not on the basis of explicit formulations. In other cases, he argues against widespread exegetical opinion: he holds that for Wittgenstein rule following is 'individualistic', though of course not private, and that a practice (of following a rule) 'can perfectly well be constituted in the regular activity of a single isolated person'; that meaning isn't constitutively normative and that Wittgenstein never suggests otherwise; and, finally, that Wittgenstein's ban on philosophical theorizing does not apply to his own (Horwich's) theory of meaning, as the latter 'is nothing more than an assemblage of potentially obvious points.' Of course – he adds-even the obvious may be missed 'by someone who is looking in the wrong direction'.

According to Diego Marconi, an important reason of the current disfavor for Wittgenstein's philosophy lies in its hostility to metaphysics, which Wittgenstein accused of 'obliterat[ing] the distinction between factual and conceptual investigations' (Z §458). Wittgenstein was particularly hostile to the idea of necessary facts, that is, to the claim that there are necessary propositions which, if true, can be known to be true only a posteriori. Such a claim was first put forth by Saul Kripke and is nowadays endorsed by many philosophers. Wittgenstein's *Tractatus* argued against the viability of the notion of a necessary fact: the argument, however, heavily depends on assumptions that are integral to the *Tractatus*'s picture theory. Nevertheless, there is a simpler version of Wittgenstein's argument that is not so committed. The latter argument has been widely discussed in recent years; Marconi shows that its failure is not uncontroversial. Thus, the anti-Wittgensteinian case for necessary facts is not as strong as it is usually taken to be.

Eva Picardi's chapter has two parts. In the first part, she argues that Frege's and Wittgenstein's endorsement of the context principle does not entail that they are committed to anything resembling today's 'radical contextualism,' as defended by Charles Travis, François Recanati, and other philosophers; moreover, the later Wittgenstein's insistence on the variety of uses of what is materially one linguistic expression should be distinguished (*pace* Travis) from the contextualists' claim that any sentence can express different propositions in different circumstances. In the second part, Picardi shows that the later Wittgenstein's views on the semantic function of proper names were generally much closer to Frege's views than some have taken them to be, in spite of Wittgenstein's anti-Fregean claim that there can be no uniform account of the semantic role of proper names. Picardi's discussion is intended to rescue Frege's and Wittgenstein's account of the semantic function of proper names from oblivion inspired by the prevalence of direct reference theories: 'Frege's and Wittgenstein's description of the way we use proper names – she remarks – is not rendered obsolete by the picture sketched by causal theories of reference.'

Timothy Williamson ends the 'meaning'-Section by giving it a slightly different twist. His polemical focus is the idea of epistemological analyticity, that is, the thesis according to which necessarily, whoever understands a sentence which is an alleged candidate for being an analytical truth (e.g., 'Every vixen is a female fox') assents to it. Not only Williamson doubts that the thesis entails other theses such as that whoever understands a sentence which is an alleged candidate for being an analytical truth knows the content expressed by that sentence, or that whoever understands a sentence which is an alleged candidate for being an analytical truth is justified in assenting to it, but also puts the original thesis into question. In his opinion, understanding one such sentence has nothing to do with recognizing that sentence as true, however this recognition is conceived. In Williamson's lights, this questioning undermines various inferentialist accounts of meaning that have been defended by many *lato sensu* Wittgensteinian philosophers (e.g., Robert Brandom and Michael Dummett), insofar as those accounts conceive the acceptance of some inference rules as constituting the understanding of the logical constants, hence of the sentences in which they occur. Although Wittgenstein did not explicitly defend an epistemological conception of analyticity, his claim that grammatical sentences – his *analog on* of analytical truths – sound obvious to people understanding them surely lies in the vicinity of that conception.

In his chapter, Pasquale Frascolla sets out a general criterion according to which any tentative interpretation of the ontology of the *Tractatus* can be assessed as correct or incorrect. The criterion is based on the identification of five theses to be found in Wittgenstein's early masterpiece, concerning the nature of objects and states of affairs, which any sound interpretation is required to consistently account for. Most of the work is devoted to commenting upon some crucial sections of the *Tractatus* from which such theses are extracted. In particular, the issue of how the status of substantiality accorded to objects can be made compatible with the possibility that an object does not occur in any of the facts into which the world divides is given a strategic position. As a result of his investigations, Frascolla makes the claim that objects are to be conceived of as abstract universals, whereas states of affairs are to be taken as concrete particulars which instantiate the former. This conclusion is attained quite independently of the answer to be given to the further question as to the phenomenalistic or physicalistic framework of the Tractatus ontology: it is only the sections devoted to solipsism which settle the matter in favor of the first interpretative option.

Wittgensteinian scholars notoriously do not agree on the weight to be attributed to Wittgenstein's explicitly metaphilosophical remarks. In her chapter, Meredith Williams makes a distinction between two parties: the party of those who maintain that, in spite of his own pronouncements, the later Wittgenstein did in fact develop theories of language and mind which in many respects are alternative to traditional philosophical theories but which, so to speak, are placed in the same space of the latter; and the party of those who take Wittgenstein's metaphilosophical remarks at face value and construe his statements as expressions of a radical rejection of the entire traditional philosophical attitude and project. Williams would like to show how to avoid the Scylla of the constructivist interpretations and the Charybdis of the quietist interpretations: in her opinion, the two rival views are undermined by the same misconception of the nature of philosophy, based on an ungrounded assimilation of philosophical theorizing with scientific theorizing. Williams illustrates her point by examining an interpretative question, that of the coherence of Wittgenstein's metaphilosophical statements with the method he actually employs in framing his arguments, and a textual question, that of the reasons for the position which the metaphilosophical remarks have in the *Philosophical Investigations*. As for the first issue, Williams's conclusion is that there is a conflict between Wittgenstein's metaphilosophical remarks and his method of arguing: the latter is not purely descriptive but is governed by a complex methodology characterized by the employment of two main forms of argument (conflation argument form and paradox argument form). As for the second issue, the answer is that the metaphilosophical remarks occur where they do because that is the point in the *Philosophical Investigations* where the conflation arguments against grammatical mistakes make room for the paradox arguments against philosophical theories.

The volume is concluded by Anthony's Kenny chapter, which is at one and the same time an assessment of Wittgenstein's philosophy in relation to the volume's themes and a commentary *of* most of the volume's contributions. As far as mind is concerned, Kenny masterfully argues that Wittgenstein endorsed an Aristotelian metaphysics according to which mental states are to be understood in the light of abilities and the exercise of them. By giving Wittgenstein an intermediate position between Cartesianism and behaviorism, this interpretation accounts for both the criteriality of the relation between mind and behavior and the inadequacy of a verificationist approach to sentences about mental states. Yet it also shows that Wittgenstein's claim that the task of philosophy is merely descriptive and not theoretical was not quite honored by Wittgenstein himself, even though he did practice a reminder of the obvious, by prompting us to see differences also by way of comparison. In this respect, Wittgenstein's conception of meaning is enlightening. In his normative approach Wittgenstein distances himself from a crypto-nomological account of normativity as the one Horwich appears to defend. Moreover, he only allows for non-social rule-following as derivative on a social one. By way of a comment, this shows that, if one guessed that Wittgenstein's stance on rule-following is rooted in the ordinary *use the expression* 'to follow a rule', this conjecture would only be prompted by too literal a reading of the famous sections of the *Investigations* on the descriptive practice in which good philosophy consists.

Part I
Mind

1
Verificationism and Wittgenstein's View of Mind

William Child

1

In a number of passages, Wittgenstein suggests that we can make perfectly good sense of ascriptions of thoughts and sensations that we have no means of verifying: thoughts and feelings that not only *are not* but *could not be* manifested in behaviour. But such ascriptions, he thinks, make sense only against a background of cases where we *can* tell what people are thinking and feeling. 'This is', he says, 'like the old business of rule and exception. A man can make a false move in chess; but it couldn't be true that everybody always made false moves in chess' (LPP, p. 99; cf. PI §345). Similarly, he thinks, we can make sense of the supposition that I guess correctly what someone is thinking even though he will not admit it. But it makes no sense to suppose that no-one ever admits what they are thinking; that would be 'like *everybody's always* making false moves in chess' (LPP, p. 99). But how, exactly, does Wittgenstein think that the intelligibility of cases in which someone will not or cannot manifest her thoughts depends on the existence of cases where thoughts are expressed? And how convincing a case does he make for his view?

Part 2 introduces Wittgenstein's discussion of cases where we understand ascriptions of thoughts and sensations that we have no means of verifying. Part 3 distinguishes different ways in which the intelligibility or existence of such cases is said by Wittgenstein to depend on the intelligibility or existence of cases in which we can tell what someone is thinking. Parts 4 and 5 consider the relation between these different kinds of dependence and discuss Wittgenstein's treatment of similar dependence relations in some other cases. Part 6 distinguishes three different accounts of the relation between our grasp of the concept of thought as it applies in cases where we can tell what someone is thinking and our grasp of the concept as it applies in cases where we cannot. I argue that there are signs of each of these three accounts in Wittgenstein; that Wittgenstein himself did not clearly favour one of these accounts over the others; but that, as a philosophical view, one of the accounts is definitely preferable to the other two.

2

Wittgenstein writes:

> The criteria for the truth of the *confession* that I thought such-and-such are not the criteria for a true *description* of a process. And the importance of the true confession does not reside in its being a correct and certain report of a process. It resides rather in the special consequences which can be drawn from a confession whose truth is guaranteed by the special criteria of truthfulness.
>
> (Assuming that dreams can yield important information about the dreamer, what yielded the information would be truthful accounts of dreams. The question whether the dreamer's memory deceives him when he reports the dream after waking cannot arise, unless indeed we introduce a completely new criterion for the report's 'agreeing' with the dream, a criterion which gives us a concept of 'truth' as distinct from 'truthfulness' here.) (PI, II, xi, pp. 222–3)

That suggests the following view. We have no concept of what it is for a subject's report of what she has just thought to be *true* other than by reference to the *truthfulness* of the report. So all there is to its being true that I was just thinking that it is Friday is that that is what I am sincerely inclined to report that I was thinking. Similarly, all there is to its being true that I dreamed that I was the Prime Minister is that, when I wake up, that is what I am sincerely inclined to report that I dreamed. As Wittgenstein puts it:

> one [cannot] ask 'How do I know that this is what I actually dreamed?' – It is there in it because I *say* it is. Or better: because I am inclined to say... (RPP, I §363)

Of course people often have dreams that they never report. But Wittgenstein's view can easily accommodate that fact: a case where someone dreams such-and-such and never reports it is a case where she is 'inclined' to say that she dreamed such-and-such but never acts on that inclination.

On the face of it, however, we can make sense of cases in which someone dreamed something but is not in any serious sense inclined or disposed to report having dreamed it. Similarly, there are cases where someone entertains a thought but is not in any serious sense inclined to report the thought. How do we make sense of these apparent possibilities? Wittgenstein himself draws attention to just such a case:

> Lytton Strachey writes that as Queen Victoria lay dying she 'may have thought of', say, her mother's youth, her own youth, Prince Albert in a Grenadier's uniform. (LPP, p. 274. See also LPP, pp. 32–3, 99, 152, 229; RPP, I §366)

As Wittgenstein says, we clearly understand Strachey's speculation. But it is perfectly possible not only that Queen Victoria did not report her dying thoughts, but that, in the circumstances, she could not possibly have done so. And in that case, we cannot make sense of claims about what she was thinking when she died in terms of what she was inclined or disposed to report thinking; in the circumstances we are imagining, she had no such disposition. So how do we understand what Strachey says?

One idea would be to appeal to counterfactuals. On this view, we make sense of the claim that as she lay dying Queen Victoria thought about Prince Albert in terms of a counterfactual: if she had been able and willing to report what she was thinking, she would have reported thinking about Prince Albert. But such a view seems implausible. And Wittgenstein himself takes a different line. His idea is that we learn 'She thought X' in connection with cases where people say what they thought, and where the question what they thought has some practical importance. But once our understanding is secure in those basic cases, he thinks, we can go on to apply the same words to cases where there is no possibility of verification, and where no practical consequences attach to someone's having thought one thing or another. He puts it like this:

> I can make conjectures about what Smith thinks and I can ask him if he is thinking that: 'I'll bet you're thinking it's about time he arrived.' This has its own kind of importance; a language-game of its own ... After someone tells you how he felt, or what he thought, you draw conclusions about what he'll do ... At some time we learn the game of answering to: 'What did you think?' I want you to regard

> telling what you thought as a completely new game. You learned it in cases of practical importance. (LPP, p. 274)

That is the basic case. And here is what Wittgenstein says about the extension to the case of thoughts for which there is no possible verification:

> Now suppose we consider: What did Queen Victoria think as she lay dying? There is no verification *accessible* to 'What did Queen Victoria think?' Then shall we say that Strachey was guessing at what she thought? You may. But it's a different use of 'guess' from the use we learned; another game... You read 'The Queen may have thought...' You think 'Yes, I understand that': this means, perhaps, 'the words have other uses'. (Cf. 'I came to a booking office': this presupposes the whole railway system: it gets its place and use from that). (LPP, p. 274)

Later on in the lectures, Wittgenstein returns to the same topic:

> We understand 'He thought X but would not admit it', but we get the use of 'He thought X' from 'He admits X', i.e. says X, writes in his diary X, acts in an X-like way ... Thinking and not admitting comes in only after thinking and admitting. It's an exception-concept. You'd have to explain to someone who did not know what 'thinking and not admitting' was in terms of thinking and admitting. (LPP, p. 329)

Wittgenstein makes similar comments about a different case in his notes from 1949–50:

> Someone groans under anaesthesia or in sleep. I am asked 'Is he in pain?' I shrug my shoulders or say 'I don't know whether he's in pain'. Sometimes I acknowledge something as a criterion for it, but sometimes I don't.
>
> Well, do I mean nothing by this? Oh yes: I am making a move in an existing game. But this game wouldn't exist if there weren't criteria in other cases...
>
> Something turns out to be pain or pretence. And that is essential to the concepts 'pain' and 'pretence', even if it is not evident in every single one of their applications. (LW II, p. 57)

Wittgenstein's idea is that our grasp of the concept *pain* is anchored in cases in which there are behavioural manifestations of pain: cases where we can tell that someone is in pain on the basis of what they say and do. Those cases are basic in determining the content of the concept. But once the concept is determined in that way, we can go on to apply it in other cases where the ordinary ways of telling that someone is in pain are unavailable: cases where there may be no way at all of telling whether or not the person is in pain.

3

I want to highlight two issues that arise from Wittgenstein's treatment of cases in which we can make sense of ascriptions of thoughts and sensations that we have no means of verifying.

First, in the cases I have mentioned, Wittgenstein distinguishes between what we might call the basic, or central cases of the application of a concept and the less basic, or derivative cases. And his formulations suggest at least two different and nonequivalent claims about the relation between the basic and the less basic cases. First, there is the suggestion that we can only *acquire* the concept as it applies in the less basic cases by first grasping its application in the basic cases; the concept of the less basic cases, we might say, is *cognitively dependent* on the concept of the basic cases. Second, there is the suggestion that the *content* of the concept in the less basic cases is essentially dependent on its content in the basic cases; the concept as it applies in the less basic case is *definitionally* or *conceptually* dependent on the concept as it applies in the basic cases. And each kind of dependence is one-way: the less basic cases depend on the basic cases, but not vice versa.[1]

Now a similar general structure can be found in Wittgenstein's treatment of a range of other concepts. There is, for example, the relation between the concepts of calculating and of calculating in the head: 'Only if you have learned to calculate – on paper or out loud – can you be made to grasp, by means of this concept, what calculating in the head is' (PI, II, xi, p. 216). There is the relation between the concepts of a correct and a false move in a game: 'A man can make a false move in chess; but it couldn't be true that everybody always made false moves in chess' (LPP, p. 99; cf. PI §345). There is the relation between the concepts *red* and *seems red*: 'When a child learns colour-names it learns first "This is red" and not "This seems red to me". It couldn't learn the latter first. The language game that uses "I see red" or "This seems red" depends on our knowing how to say "This is red"' (LPP, p. 32; cf. Z §§418 ff). And there is the relation between the application of the concepts *fat* and *lean* to people and their application to days of the week (see PI, II, xi, p. 216).

The words get their primary sense in connection with their application to people. But once they have that sense, they can be applied in a 'secondary sense' to days of the week.

What is the relation between the different dependence-claims that Wittgenstein makes? What, if anything, is supposed to justify or explain them? And are Wittgenstein's claims defensible?

A second topic that emerges from Wittgenstein's treatment of the Queen Victoria case and others like it concerns verification and verificationism. The idea that there is an important relation of some kind between the way a proposition is verified or justified and its meaning is a consistent thread in Wittgenstein's writings and lectures from 1929 until his death. His view about exactly what the relation is varies over time. But he never gives up the idea that there *is* an important relation between meaning and verification. Thus, for example, he writes in *Philosophical Investigations*:

> Asking whether and how a proposition can be verified is only a particular way of asking 'How d'you mean?' The answer is a contribution to the grammar of the proposition. (PI §353)

Similarly, there are a number of passages from 1946–8 where he argues that the concept of length is not independent of the method of judging length and, relatedly, that the concept of the motive for an action is not independent of our methods of judging motives (see PI II, xi, pp. 224–5, RPP, I §§631–2, LPP, pp. 263–4). And so on. But how close is the relation between meaning and verification supposed to be? In *Philosophical Remarks*, Wittgenstein thought the relation was very close: 'How a proposition is verified is what it says', he wrote (PR §166); 'It isn't possible to believe something for which you cannot imagine some kind of verification' (PR §59). Part of the point of Wittgenstein's discussion of the Queen Victoria case is to acknowledge that the relation between meaning and verification is weaker than that. The example is explicitly introduced as one in which a proposition makes sense even though, in the nature of the case, it is not 'accessible to verification' (cf. LPP, pp. 152, 274).[2] And the remarks about pain quoted earlier from LW II, p. 57 are similar: their point is that I can make sense of the claim that a sleeping or anaesthetised person is feeling pain, even if I have no idea at all about how one could go about telling whether the person is in pain or not. Exploring these cases casts light on the way Wittgenstein saw the relation between verification and meaning.

4

Wittgenstein claims that one cannot grasp the application of the concept *S thinks that p* in cases where S will not admit thinking that p without first grasping its application in cases where S does admit what she thought. But why? What explains the cognitive dependence of cases of one kind on cases of the other?

It might be suggested that the dependence is a matter of empirical fact: if a child is brought up in an environment in which people never say what they are thinking, she will never be able to formulate the concept of thought. As an empirical claim, it is entirely plausible that such a child would never learn to conceive of others, or herself, in ways that employ the normal range of psychological concepts. But it is one thing to say from the armchair that that is a plausible claim. It is another thing to show that it is true. And even if it is true that children would not in fact develop the concept of thinking in such circumstances, that does not show that it would be impossible for them to do so. But Wittgenstein wants to show that it does not *make sense* to think of someone understanding applications of the concept of thought in cases where thoughts are not manifested without first understanding applications of the concept in more basic cases, where it is possible to tell what someone is thinking. And no merely empirical argument could establish that conclusion.

When Wittgenstein makes claims of cognitive dependence, he is not interested in the *process by which* we come to acquire the concept of the less basic case. His focus is, rather, on the question, *what it takes* for the concept we acquire to be the concept it is. That point is well illustrated in his discussion of the concept of dreaming. He offers the following suggestion about how the concept might be acquired:

> people narrate something to us after waking up; we thereupon teach them the expression 'I dreamed...' followed by the narrative. I then sometimes ask them: 'Did you dream anything last night?' and sometimes get an affirmative, sometimes a negative answer, sometimes a dream narrative, sometimes none. That is the language-game. (RPP I, §368; cf. PI, II, vii, p. 184)

But he goes on to make clear that the way in which the concept is acquired is not essential to its being the concept it is:

> There is also this to say: the child does not absolutely have to learn the use of the word 'to dream' by first merely reporting an occurrence

> on waking up, and our then teaching it the words 'I dreamed'. For it is also possible that the child hears the grown-up say he has dreamed and now says the same thing of itself and tells a dream. I am *not* saying: the child *guesses* what the grown-up means. Suffice it that one day it uses the word, and uses it under the circumstances under which we use it.
>
> So the proper question is not: 'How does he learn the use of the word?' but rather 'how does it come out that he does use it as we do?'. (RPP I, §§375–6; cf. RPP II, §241)

Similarly, Wittgenstein holds that 'you can only learn what "calculating in the head" is by learning what "calculating" is' (PI, II, xi, p. 220). But the important question, for him, is not how we acquire the concept of calculating in the head but, rather, what it takes for the concept we acquire to be the concept of calculating in the head. So his reason for saying that one could not *acquire* the concept of calculating in the head without first having the concept of calculating out loud or on paper is not that it would be too difficult; it is, rather, that nothing would *count as* the concept of calculating in the head if one did not relate it appropriately to the concept of calculating *simpliciter.* So the fact that the concept of calculating in the head is *cognitively* dependent on the concept of calculating *simpliciter* is explained by the fact that the first concept is *conceptually* or *definitionally* dependent on the second. And, for Wittgenstein, the same point applies across the board. If we want to understand why one concept, or kind of application of a concept, is cognitively dependent on another, we must ask why the first concept, or kind of application, is conceptually dependent on the second.

That, at any rate, is Wittgenstein's view. But it is not the only possible view. One might, for example, hold that it is possible for the relations of cognitive dependence and conceptual dependence to run in opposite directions. On this view, there can be concepts A and B for which the following is true: concept A can only be explained or defined in terms of concept B; but no-one can acquire concept B without first having concept A. That is how Christopher Peacocke sees the relation between the concept *red* (our ordinary colour concept, which applies to external objects) and the concept *red'* (a concept that applies to regions of the visual field) (Peacocke, 1983, pp. 42–3). The concept *red,* Peacocke thinks, is conceptually or definitionally dependent on the concept *red'*; what it is for an object to be red is for it to be disposed to present itself in a red' region of the visual field. But the concept *red'* is cognitively dependent on the concept *red*; no-one can acquire the concept *red'* without first having the concept *red*. I will not here pursue the question whether Wittgenstein is right to suggest that cognitive dependence is simply a reflection of conceptual dependence. But it is important to note both that that is his view, and that it is not the only view one could take.

5

Wittgenstein thinks the concept of calculating in the head essentially depends on the concept of calculating. But what explains that fact (if it is a fact)? And what explains or justifies the parallel claims he makes in other cases? In particular, why is the application of the concept of thinking in cases where there is no way of telling what someone thought essentially dependent on its application in cases where someone does say what she thinks?

In some of Wittgenstein's examples, it is easy to see why he thinks that one concept, or kind of application of a concept, is dependent on another. For example, Wittgenstein holds that we cannot explain the concept *looks red* except in terms of the concept *red*. And the reason for that seems straightforward. The property of looking red is the property of appearing to be a certain way. And we cannot specify *how* something has to appear in order to look red without employing the concept of the property something appears to have when it looks red. Similarly, the concept of a toy train is dependent on the concept of a real train. A toy train is something that is designed to mimic or represent a train. So we cannot specify what it is for something to be a toy train without using the concept of a train.

A different rationale is involved in Wittgenstein's claim that 'a man can make a false move in chess; but it couldn't be true that everybody always made false moves in chess' (LPP, p. 99); 'it has no meaning to say that a game has always been played wrong' (OC §496). The idea here is that what it is for something to be a legitimate move in chess is simply for it to be a move that we all ordinarily count as legitimate; so it could not be the case that the moves that we all ordinarily make – the moves that we all ordinarily count as legitimate – were actually false moves.

Another kind of case is illustrated in Wittgenstein's example of the application of the words 'fat' and 'lean' to days of the week:

> Given the two ideas 'fat' and 'lean', would you be rather inclined to say that Wednesday was fat and Tuesday lean, or *vice versa*? (I incline decisively towards the former.) Now have 'fat' and 'lean' some different

meaning here from their usual one? – They have a different use. – So ought I really to have used different words? Certainly not that. – I want to use *these* words (with their familiar meanings) *here*...

Asked 'What do you really mean here by "fat" and "lean"?' – I could only explain the meanings in the usual way. I could *not* point to the examples of Tuesday and Wednesday.

Here one might speak of a 'primary' and 'secondary' sense of a word. It is only if the word has the primary sense for you that you use it in the secondary one. (PI, II, xi, p. 216)

The words 'fat' and 'lean' get their meanings in connection with the basic or 'primary' cases. Once they have those meanings, we can go on to use them in connection with the 'secondary' cases. And the sense of the words in these latter cases, their 'secondary' sense, is essentially dependent on their sense in the primary cases: we apply the word 'fat' *with its familiar meaning* to a day of the week. But we do not suppose that the word literally applies to a day in the same sense that it applies to a person or animal. Here again, Wittgenstein's reason for thinking that the meaning of a word in the secondary cases depends on the meaning it has in the primary cases seems clear.[3]

What about the dependence of the concept *calculating in the head* on the concept *calculating*? The remark, 'Only if you have learned to calculate – on paper or out loud – can you be made to grasp, by means of this concept, what calculating in the head is' (PI II, xi, p. 216), comes immediately after the passage about primary and secondary sense that I quoted above. But it would be uncharitable to suppose that Wittgenstein thinks the application of the concept *'calculating'* to cases of calculating in the head is exactly analogous to the application of the word 'fat' to days of the week. The point is just that there is a similar relation of dependence. And we might explain the dependence in the following way. What we learn to call 'calculating', Wittgenstein thinks, is the procedure of calculating out loud or on paper. Calculating in the head is then defined as something that resembles calculating in one respect (one starts with a numerical problem and finishes with a solution) but differs from it in another (there is no overt procedure with spoken or written numerals). That is part of Wittgenstein's reason for thinking that the concept *calculating in the head* cannot be explained except in terms of the concept *calculating*. But there is a second reason, too. Imagine a group of people who do something that, considered in itself, is just like what we do when we calculate in our heads, but who know nothing of any practice of calculating out loud or on paper. *We* can see what they do as being relevantly similar to cases of calculating *simpliciter*. But *they* cannot see it that way. Now consider the question: Is what they do calculating in the head? If we do count this as calculating in the head, Wittgenstein thinks, we must depict it as 'a limiting case' (PI §385). That implies that a clear case of calculating in the head requires not just that the subject does something that is relevantly similar to calculating out loud or on paper, but also that she thinks of what she does as being relevantly similar to these basic cases of calculating. So the subject herself must have the concept of calculating *simpliciter*.[4] That is a second way in which the concept of calculating figures in explaining the concept of calculating in the head.

6

The cases we have been considering illustrate a number of different ways in which one concept, or kind of application of a concept, may depend on another. With these cases in mind, we can turn to the concept of thinking. Why does Wittgenstein hold that the application of the concept *thinking* in the Queen Victoria case depends on its application in cases where people say what they thought? What would be wrong with the idea that there is no special connection between the content of the concept *thinking* and the case where thoughts are manifested, that the Queen Victoria case is just as basic or paradigmatic a case of thinking as any other, and that there is nothing secondary or dependent about the application of the concept of thought in such cases?

Some applications of mental concepts clearly are 'secondary' or derivative. Thus, Wittgenstein writes:

> We do indeed say of an inanimate thing that it is in pain: when playing with dolls for example. But this use of the concept of pain is a secondary one. (PI §282)

When we say that a doll is in pain, we are pretending that it is a real child and pretending that it is in pain. So the content of the pretence can only be understood by reference to the primary use of the concept *pain*, in thinking about the pains of real people. That is why the application of the concept *pain* to dolls is entirely dependent on its application to people and animals. But Wittgenstein presumably does not think that the application of the concept of thought in Queen Victoria cases is secondary in *that* way. When we speculate about Queen Victoria's dying thoughts, we are not merely *pretending* that she thought this or that: we think our speculations about what she thought are literally true or false. So what other model does Wittgenstein have in mind when he says that the application of the concept of thought in the Queen Victoria case depends on its application in cases where thoughts are manifested?

Wittgenstein evidently thinks that a central role is played, in determining the content of the concept of thought, by cases in which what someone is thinking is manifest in their words or actions. We could regard that thought as an application of a more general principle: that a central role is played in determining the content of a concept by cases in which the concept is manifestly instantiated. That general principle does not apply to every concept. The content of a highly theoretical concept, for instance, is determined by the theory in which it appears, not by cases where it is manifestly instantiated; indeed, there need be no cases in which it is manifestly instantiated. Similarly for concepts that can be analysed in terms of simple descriptive conditions. If I know that a spy is a person whose job is to collect and report secret information on an enemy or competitor, then I have the concept of a spy. I need have no way of telling whether someone is a spy; and no part is played in determining the content of the concept *spy* by cases in which someone is recognisably a spy.[5] It is very plausible, however, that there are some cases where the principle does apply.

Colour concepts are an obvious example. Cases in which something is manifestly red, in which it is observed to be red, have a crucial role in determining the content of the concept *red*. But the concept *red* can also be applied to things that are not observed to be red, and even to things that in some reasonably strong sense *could not* be observed to be red: things that can only exist in conditions where human life is impossible, and so on. How should we understand the application of the concept in those cases? One idea is to appeal to counterfactuals: for an unobserved object to be red is for it to be true that, if it *were* observed by a normal observer in normal conditions, it *would* look red. That formulation might work in explaining how we make sense of applications of the concept *red* to objects that merely *are not* observed. But it is hard to see how it could work for the case of an object that *could not* be observed to be red. Yet it does seem possible to make sense of the thought that such an object is red. So how do we make sense of it? An obvious proposal is this: Cases in which objects are manifestly red play an essential role in determining the content of the concept *red*. Once the content of the concept is fixed in that way, we can then explain what it is for an unobserved object to be red by appeal to an identity: an unobserved object is red just in case it is *the same colour as* an object that is observed to be red. On this view, the application of the concept *red* to unobserved and unobservable things is cognitively and conceptually dependent on its application to observed things. But the sense in which an unobserved or unobservable object is red is exactly the same as the sense in which an observed object is red: there is nothing secondary, or nonliteral, or extended about the way in which the concept *red* applies to unobserved and unobservable objects.[6]

Now it might be said that such a view is inadequate or question-begging, for essentially Wittgensteinian reasons. For, the thought would go, if we are trying to explain what it is for an unobserved object to be red, we cannot simply help ourselves to the idea of the object's being *the same colour* as an observed red object. Adapting what Wittgenstein famously says about a different case: We know well enough that one can call an observed red thing and an unobserved red thing 'the same colour', but what we do not know is in what cases one is to speak of an observed and an unobserved thing 'being the same colour' (cf. PI §350). So we need to give some more substantial account of *what it takes* for an unobserved thing to be the same colour as an observed red thing. And the present proposal gives no account of that at all.

But how far would Wittgenstein push this objection? He would certainly insist that what it takes for one thing to be the same colour as another cannot just be taken for granted: it must be understood by reference to our concept of colour; and that concept involves, or depends on, a whole practice of sorting and classifying things according to their colours, of identifying and talking about colours, of comparing colours, of agreeing and disagreeing about which things are the same colours, and so on. That much about Wittgenstein's view is clear. But once that point is taken on board, does Wittgenstein think there is a further problem about extending the concept *red* from things that are observed to things that are not observed, and to things that could not be observed? For the case of objects that are not, but could be, observed, there are indications that he would take the dispositional route mentioned above, and say that what it is for an unobserved thing to be red is for it to be disposed to appear red to the normal sighted under certain circumstances.[7] But what about applying the concept *red* to things that *could not be* observed? I know no passage where Wittgenstein explicitly considers that question.[8] Perhaps he would regard such an application as unintelligible. But if that is his view, it needs further argument to show why we should accept it. On the face of it, we could accept Wittgenstein's view that what counts as one thing's being the same colour as another is determined by an actual practice of classifying colours while accepting that it is perfectly intelligible to suppose that an object may be red even if it could not possibly be observed to be red. There is no obvious reason why the concept of colour that we develop in connection with practices involving observed things should not straightforwardly apply to things whose colours we could not observe.

What about the concept of thinking? Two points about Wittgenstein's view are clear. First that, as I have said, a central role is played in determining the content of the concept by cases in which we can tell what someone is thinking because she says or otherwise manifests what she is thinking. Second, that our grasp of what it is for someone to think so-and-so in a case where her thoughts cannot be manifested is dependent on our grasp of what it is for someone to think so-and-so in a case where her thoughts are manifested. But exactly what is the relation between the content of the concept in the two kinds of case? We can distinguish three quite different models, each of which is consistent with the two points just made.

On the first model, the relation between the case where someone says what she is thinking and the Queen Victoria case is like the relation suggested earlier between the cases of observed colour and unobservable colour. Cases in which we can tell what someone is thinking play an ineliminable role in acquiring and explaining the concept of thinking. But the concept we acquire in connection with those cases can be applied without further explanation to cases where people's thoughts are not and could not be manifested; the content of the concept of thinking as applied in the Queen Victoria case can simply be read off the concept as it applies in the case where someone says what she thinks. On this view, explaining how we understand the sentence, 'As she lay dying, Queen Victoria thought about Prince Albert,' does not require us to say anything beyond what we say in explaining the concept of thinking as it applies in cases where we can tell what someone is thinking. At one point in his discussion, Wittgenstein presses the question, how we use that sentence. He writes:

> Don't we *understand* it, when Strachey makes surmises about what Queen Victoria may have seen in her mind's eye just before her death? Of course – but didn't people also *understand* the question how many souls there was room for on the point of a needle? That is to say: the question whether one understands this does not help us here; we must ask *what* we can do with such a sentence. – *That* we use the sentence is clear; *how* we use it is the question. (RPP I, §366)

On this first model, that question has a straightforward answer. We use the sentence 'Queen Victoria saw so-and-so before her mind's eye' to speculate about Queen Victoria's dying thoughts. We engage in such speculation because we are interested in what she was thinking about immediately before her death. And we are interested in that question for its own sake – not because we think it has any practical implications.

It may be that Wittgenstein would accept that answer, and the model of our understanding of the Queen Victoria case that goes with it. But some of what he says suggests that he would regard the answer as inadequate, and would favour a quite different model of our understanding of the case. On this second model, the content of the concept of thought as applied in the Queen Victoria case cannot simply be read off the content of the concept in the more basic cases; it must be understood by giving a direct account of the nature and point of the practice of describing and speculating about thoughts whose ascription cannot possibly be verified. We find it natural to take the word 'thought' from the basic cases, where we can tell what someone is thinking, and apply it in Queen Victoria cases. The meaning of the word in these new applications is parasitic on its meaning in the basic cases. But it is not fully determined by that use; it depends also on the actual use of the word in the new applications. So, in using the word 'thought' in these new cases, we are *extending* its sense: *giving it* an application to the Queen Victoria cases that it did not already have. This new use depends on our shared interest in developing narratives about the inner lives of others: narratives that have no practical purpose, and for which there is no standard of correctness other than what people agree in regarding as plausible or appropriate. On this view, the practice of discussing Queen Victoria's dying thoughts comes closer to the practice of discussing fiction than to that of ascribing thoughts in more basic cases.[9]

A third model is suggested by the following passage:

> What is the purpose of a sentence saying: perhaps N had the experience E but never gave any sign of it? Well, it is at any rate possible to think of an application for the sentence. Suppose, for example, that a trace of the experience were to be found in the brain, and then we say it has turned out that before his death he had thought or seen such and such etc. Such an application might be held to be artificial or far-fetched; but it is important that it is *possible*. (RPP I, §157)

That suggests the following view: The sentence 'perhaps N had the experience E but never gave any sign of it' has an application, and therefore a meaning. But it has that meaning only in virtue of the fact that there is in principle some way of verifying whether or not N did have the experience E: a method that goes via physical traces rather than behavioural expressions. If we apply this line to the Queen Victoria case, we will say that we understand the ascription of thoughts in such a case by supposing that there is in principle some way of verifying such ascriptions: a method that looks to physical traces of her thoughts. I noted earlier that Wittgenstein introduces the Queen Victoria case as one in which we can make sense of ascriptions of thoughts for which there seems to be no possible verification. On this third model, it turns out that the initial appearance of unverifiability was misleading: Queen Victoria's dying thoughts cannot be verified by reference to her words or actions; but they can at least in principle by verified by reference to what is going on in her brain; and it is only in virtue of that fact that one can make sense of ascribing such thoughts at all. It seems fair to regard this third view as a form of verificationism.

If Wittgenstein accepts the first model of our understanding of the ascription of thoughts in the Queen Victoria case, his treatment will be decisively nonverificationist. If he accepts the second model, his account will not be a form of verificationism; but it will certainly be a form of antirealism. For it will explain the meanings of these ascriptions in a way that gives up the idea that there is a straightforward fact of the matter about what Queen Victoria was thinking in her dying moments. If Wittgenstein accepts the third model, then his account of the Queen Victoria case is, after all, a form of verificationism. For the meaningfulness of ascriptions of thought in the Queen Victoria case will depend on the supposition that those ascriptions are not, after all, inaccessible to *every* form of verification.

Which of the three models would Wittgenstein accept? I think his position is unclear. The first model is consistent with much that he wants to say. But there is some evidence that he would reject that model; that he would insist that an account of the meaning of the word 'think' as applied in Queen Victoria cases must say something more substantive about our practice of using the word in such cases. The very fact that he presses the question, what we *do* with the sentence 'Queen Victoria may have thought such-and-such' suggests that, even when we have explained the meaning of ascriptions of thought in cases where a subject's thoughts are manifested, there is a further question, how we understand ascriptions of thoughts that lie beyond our normal methods of verification. That, in turn, suggests that the concept of thought that we grasp in connection with the basic case where people manifest their thoughts cannot be applied without further explanation to cases where a person's thoughts cannot be verified in our normal ways; and it suggests that, when we apply the concept of thought in Queen Victoria cases, we are in some way developing or extending the concept. A view of that sort seems right for the application of the adjectives 'fat' and 'lean' to days of the week. Perhaps it is right for the application of the concept *calculating* to cases in which there is no overt process of calculation. But it is hard to believe that it is right for the application of the concept *thinking* to Queen Victoria cases. If Wittgenstein was indeed tempted by such a view, it is a temptation he should have resisted.[10]

Notes

1. The terminology used here draws on the discussion of different kinds of dependence in Peacocke (1983, pp. 42 ff.).
2. For a different reading of Wittgenstein's discussion of the Queen Victoria case, see Forster (2004, pp. 172–82), who regards Wittgenstein's treatment of the case as exemplifying a clearly verificationist strand in his late philosophy of mind. Forster stresses a passage in LPP that, he thinks, implies that claims about completely unverifiable thoughts are senseless: 'If [Strachey] could not conceivably know [what Queen Victoria was thinking], what sense could this have? If no sense, then why say it?' (LPP, p. 229). But it is wrong to read that passage as showing that Wittgenstein thought that, if Strachey's speculations are unverifiable, then they are senseless. In the first place, Wittgenstein is asking questions, not making statements. In the second place, it is not at all obvious that the answer to Wittgenstein's first question is supposed to be: 'It could have no sense'. On the contrary, it seems plain that his starting point is that Strachey's speculation *does* make sense; the task is to see how it makes sense. (That is explicit in RPP, I §366)
3. For a related case, see RPP I, §125: 'The feeling of the unreality of one's surroundings. This feeling I have had once, and many have it before the onset of mental illness.... But why do I choose precisely the word "unreality" to express it? Surely not because of its sound. (A word of very like sound but different meaning would not do.) I choose it because of its meaning. But I surely did not learn to use the word to mean: a *feeling*. No; but I learned to use it with a particular meaning and now I use it spontaneously like *this*'.
4. Compare the following from PI §282: 'When children play at trains their game is connected with their knowledge of trains. It would nevertheless be possible for the children of a tribe unacquainted with trains to learn this game from others, and to play it without knowing that it was copied from anything. One might say that the game did not make the same *sense* to them as to us.'
5. I owe this example to Timothy Williamson.
6. This model of the relation between the basic case and the less-basic case draws heavily on Peacocke's account of 'identity-involving explanations of concept possession' in Peacocke (2008; see especially chapter 5, 'Concepts of Conscious States').

7. See RC III §97: 'Don't we just *call* brown the table which under certain circumstances appears brown to the normal sighted?'
8. PI §§514–15 considers the question, whether a rose is red in the dark, in the context of a discussion of forms of words that look like intelligible sentences but are not (see also Z §250). But what Wittgenstein is arguing in that passage is not that the claim that a rose is red in the dark is unintelligible but, rather, that it is not the possibility or impossibility of imagining a rose being red in the dark that shows whether or not the sentence is intelligible.
9. It is interesting to note that, in LPP, the question why we are interested in fiction is raised immediately after a discussion of the Queen Victoria case (LPP, pp. 32, 152, 274).
10. Earlier versions of some of this material were presented at the conference 'Is there anything wrong with Wittgenstein?' in Reggio Emilia, June 2006; at a workshop in honour of Gunnar Svensson, 'Themes from Wittgenstein', at Stockholm University, February 2008; at the International Wittgenstein Symposium in Kirchberg, August 2008; and at the Universities of York and Nottingham. I am grateful to the participants on those occasions, and to the editors of the present volume, for many helpful suggestions.

2

Reading-Machines, Feelings of Influence, Experiences of Being Guided: Wittgenstein on Reading

Joachim Schulte

The book *Philosophical Investigations* as we know it was not written in one go. The first part of the text (up to the beginning of §189) was originally composed in Norway, between November 1936 and the summer of 1937. Roughly the second part was written in 1944, and the third part (which is very much a collage of early and late material) was assembled and probably completed in 1946. The section on reading (as I shall call it) comprises §§156 to 178, that is, more than 11 pages of the text of the English edition of the book. The history of this part of the text is different from any other part, and in contrast to many other parts its history is easily told. When in 1936–7 Wittgenstein composed the manuscript version (the *Urfassung*[1]) of what became *Philosophical Investigations* he lifted the entire section on reading from his aborted attempt at producing a German translation and revision of the *Brown Book* and inserted it into his new manuscript, the *Urfassung* of the *Investigations*.[2] In doing so he made comparatively few changes, leaving out a few remarks, adding one and changing some words. One is tempted to say that the biggest change consists in prefixing the remarks with numbers.

The German *Brown Book* revision is the manuscript which Wittgenstein had been working on in his Norwegian cabin immediately before embarking on the *Urfassung* of the *Investigations*. We know that he was dissatisfied with this revision, for at a certain point he abruptly stopped working on it, commenting that the whole work was worth nothing. It is in accordance with his dissatisfaction that he did not use much material from the German *Brown Book* revision when working on the manuscript of the *Urfassung* – with the single exception of the entire section on reading. The fact that he simply inserted this section into his new manuscript and refrained from making many changes in it at later stages of the composition of the *Investigations* is good evidence for assuming that he was fairly satisfied with this material.

There is one further stage in the history of this section on reading that should be mentioned. Great parts of the German *Brown Book* revision are a free translation of the original *Brown Book* plus a number of minor additions and minus a number of minor deletions. The section on reading is different. The part corresponding to §§156 to 164 of the *Investigations* is basically a translation of the *Brown Book* material. The rest, stretching from §165 to §178 (that is, roughly the second half of the entire section), was added in the course of producing the German revision more than a year after Wittgenstein had dictated these thoughts to Francis Skinner and Alice Ambrose. The (much shorter) *Brown Book* section comes at a particular point of that dictation, viz., just before the end of Part I of the *Brown Book*. I take it that the division of the *Brown Book* into two parts is due to purely external circumstances. Probably work was interrupted because of the vacation between terms. So the early version of the section on reading (up to PI §164) concludes the considerations of Part I, except for a fairly general remark about the family resemblances between different phenomena of reading or being guided and a kind of summary, stating that the examples given were not meant to show us the essence of reading 'through a veil of inessential features', as he says (BB, p. 125). His discussion of these examples is not an *indirect* means of indicating something that cannot be shown: the examples do not serve as hints at something hidden or lost. In the German revision of the *Brown Book*, he rather sweetly writes that this indirect method would have to be used if he wanted to convey what it had been like in his youth when people danced the waltz. Not so in the cases of reading, being guided, and so on. And he concludes this part of the *Brown Book* by insisting that 'Our method is *purely descriptive*; the descriptions we give are not hints of explanations' (*ibid.*).

In the observations that follow I propose to treat the section on reading as a kind of separate chapter of the book *Philosophical Investigations*. This may easily appear to be an unusual approach. I have gained the impression that this section does not have many admirers. Some people seem to treat it as a kind of feeble anticipation of the famous remarks on rule-following – a view that I find slightly baffling.[3] Other readers treat it as a kind of digression, a somewhat bewildering attempt at elucidating the remarks on understanding that precede it, providing a loose connection with subsequent remarks. To be sure, the first remark of our section refers back to what went on before and claims that this (viz., what was said before) will become clearer through the following remarks. But that, of course, is a pretty general statement which does not imply anything about a possibly subordinate or prevalently ancillary status of the section on reading. Wittgenstein himself uses the word '*einschalten*' (interpolate), which suggests that his following remarks are meant to be seen, not as a direct continuation of the preceding material, but as a chapter of its own. I want to treat it as such and see how far I can get that way.

I think that one of the main difficulties readers have encountered in dealing with the section on reading is a lack of clarity about the real drift of Wittgenstein's words. Can his remarks be seen as contributing to an identifiable point? Are they directed at a recognisable target? Do they serve to explain a specific point of view? These and similar questions can be taken to express a certain perplexity about the thrust of these considerations. I sympathise and have to admit that I feel far from sure about the right or most profitable way of reading this section. On the other hand, I suppose that the facts I have mentioned about the history of this section indicate that Wittgenstein himself was more or less satisfied with these observations. And in view of the extremely self-critical attitude he normally adopted towards his writings this should be reason enough to look for a reading that could help to explain his satisfaction.

My title is meant to highlight three of the most striking topics of Wittgenstein's discussion. The first topic is that of reading-machines. The idea is mentioned only towards the beginning of our section, and its first occurrence does not sound terribly dramatic. Wittgenstein begins by laying down that his use of the word 'read' does not imply that the person who is described as reading something understands the sense of what he reads. Wittgenstein says that 'reading is here the activity of rendering out loud what is written or printed; but also of writing from dictation, copying something printed, playing from a score, and so on' (PI §156). Somewhat surprisingly, he then goes on to talk about our ordinary use of the word 'read' which, as his own description makes clear, goes beyond his minimalist stipulation and does involve a reader's grasp of the sense of what he reads. An attentive reader will be able to paraphrase what he has read, while a distracted reader will not be able to say anything about the content of the sounds he has uttered. Wittgenstein himself puts the contrast by saying that a reader:

> may attend to what he reads, or again – as we might put it – function as a mere reading-machine: I mean, read aloud and correctly without attending to what he is reading; perhaps with his attention on something quite different (so that he is unable to say what he has been reading if he is asked about it immediately afterwards). (PI §156)

This use of the expression 'reading-machine' is no more than a graphic way of characterizing an inattentive or purely mechanical manner of reading.

The next remark begins with a description of a startling thought-experiment. We are asked to imagine a situation in which human beings or other creatures are used as reading-machines and trained to read out loud whatever we want them to read. Starting from here, Wittgenstein considers a case where a pupil responds in a way which sounds roughly right while his teacher insists that this is only accidentally so. After a few more exercises, however, the pupil's performance meets with the approval of the teacher, who now agrees that the learner can read. This serves to show that there is no exact point at which we are justified in saying that *now* the pupil can read whereas the preceding word uttered by him was not yet the product of an act of genuine reading.

Well, this does not really look like an especially exciting result. In particular, why bring in the whole rigmarole of an imagined world with live reading-machines to arrive at this one insight? One feels that an examination of an ordinary teaching-and-learning situation would have led to the same result. So why does Wittgenstein bring in reading-machines at all?

Presumably the point is connected with two different sorts of cases Wittgenstein mentions in order to contrast them with the present one. The first of these cases is such that we use the word 'read' to stand for 'a certain experience of transition from marks to spoken sounds' (PI §157). Here it would indeed make sense to speak of a first time or a first word that was read by a given person. It would simply be the word with reference to which this person would say that when he pronounced it he for the first time had the feeling that he was reading. This would be parallel to a statement like 'This was the first glass of Californian red wine that I enjoyed' or 'This was the first time I actually *liked* listening to Callas.'

The second case is that of a pianola-like sort of reading-machine, where you can say that the moment certain wires were connected with such-and-such parts of the apparatus the machine could read. The first word was, for example, 'belly-dancing'.

In the case of our live reading-machines, on the other hand, the change brought about by training these pupils is to be described in terms of their responses and other aspects of their behaviour. Perhaps it would be still better to speak of *patterns* of behaviour. Changes in a creature's pattern of behaviour are gradual changes. There is never one precise point where you could say things like 'At 7:30 pm he stopped being a boor and began his career as mother's darling.' Similarly, you cannot say that a slave named Ludwig who was trained to become a live reading-machine started to *read* when ten minutes ago he pronounced the word 'deeply,' while the words 'The Mock Turtle sighed,' which he had uttered immediately before that, were not really read.

So there is a specifiable difference between different uses of the word 'read.' And these different uses can be exemplified in terms of the case of live reading-machines, on the one hand, and cases like that of pianola-like machines as well as that of reading-experiences, on the other. Now, the matter becomes interesting as soon as you wonder about how to arrange these different sorts of cases. Asked to put our cases in a natural order, surely all of us would tend to begin with the experience-man, place the live reading-machine next and end up with the pianola-like machine. Or we would do it the other way around: but at any rate, the live reading-machine would be placed in the middle between the other two sorts of case. By describing his example in a certain way Wittgenstein has managed to confuse this seemingly natural order: the experience-man and the pianola are on one side, the live reading-machine on the other.

What is more, the experience-man and the pianola (but not the live reading-machine) are placed on the side of mechanism while the live reading-machine is non-mechanical. Wittgenstein says that in this case the concept 'reading' is 'quite independent of that of a mental or other mechanism'. The pianola and its reading performance are obviously defined through its mechanism. But also in the case of the experience-focussed description of a conscious act of reading it seems most natural to read such a description in mechanistic, or quasi-mechanistic, terms. As Wittgenstein writes in a remark that comes later in our section on reading:

> We imagine that a feeling enables us to perceive, as it were, a connecting mechanism between the look of the word and the sound that we utter. For when I speak of the experiences of being influenced, of causation, of being guided, that is really supposed to mean that I, so

> to say, feel the movement of the levers which connect seeing the letters with speaking. (PI §170)

I think that the question we are meant to ask ourselves is whether the description in terms of live reading-machines (that is, in terms of responses and patterns of behaviour) would not be a more adequate model for characterizing complex human capacities than either of the two mechanistic ones. In other words, the suggestion is that the terms that are sufficient to describe the behaviour of live reading-machines may well be good enough, and perhaps even the best available, to describe what is going on in ordinary cases of human reading.

One may want to give an account of reading in the restricted, or minimalist, sense explained above that uses the notion of a derivation. Looked at in this way, reading involves making a transition from a printed letter, the original pattern, to a spoken sound. It is in this sense that you may be said to derive a sound, a spoken word or phrase, from the printed page. It appears quite easy to describe everything that is going on in such a case of derivation in the behavioural terms suited for describing the activities of live reading-machines. All that is needed is a reference to the rule of derivation, what Wittgenstein calls the 'rule of the alphabet', on the one hand, and the pupil's performance, which shows that he is using the rule, on the other. To make it clearer that the pupil's behaviour can actually *show* this, we may modify the example by focussing on the actions of a person who is told to copy printed letters by using ordinary handwriting. The rule is explicitly given by means of a table showing printed letters on one side and cursive letters on the other. This way, the idea of a derivation in accordance with a rule is unambiguously spelled out, and it is also clear that behavioural terms are sufficient to describe what is going on.

Now one might think that our example is too simple to tell us much about our concept of reading. So Wittgenstein suggests a way of complicating matters by mentioning the possibility of deriving letters in a modified order of the alphabet: you move from A to b, from B to c, and so on. This possibility is reminiscent of the code used by Wittgenstein himself and several members of his family, who used it for private notes or messages. Such a code can of course be modified by introducing ever more complicated rules of derivation until at some point the question will be raised whether it is possible to make a clear distinction between this kind of procedure and a random one (PI §163). This question, however, should not lead us to think that the word 'derive' has no real meaning, nor that the supposed meaning disintegrates under the impact of a series of examples of complicated rules. What these examples and others like them show is that 'reading' or 'deriving' are family-resemblance concepts: it is a mistake to think that you can discover the essential meaning by abstracting from individual cases. If you want to explain the meaning of such a word, you will, as Wittgenstein says, have to describe 'a selection of examples exhibiting characteristic features, some examples showing these features in exaggeration, others showing transitions, certain series of examples showing the trailing off of such features' (BB, p. 125).

This is as far as Wittgenstein got in the *Brown Book*: he made the point about family resemblance and then moved on to other topics. In the German revision of the *Brown Book*, however, and in the corresponding remarks of the *Investigations* he finds it instructive to continue the discussion of reading by considering a number of attempts at complementing the behavioural account. These attempts aim at bringing out certain features of the phenomenology of reading (as one may want to put it) and showing that an account of reading would be incomplete if it did not consider them.

Thus one may be inclined to say that a description of an act of reading is far from complete if it does not mention more than the reader's looking at printed words and pronouncing certain sounds. It is possible, after all, to look at a printed page and speak certain words without having read them in any sense of 'read'. And this may even be the case if the spoken words are, or happen to be, exactly those that you would have pronounced if you *had* read them. Of course, there is a difference between reading a word, on the one hand, and associating a word with a printed pattern of letters or letting oneself be reminded of a word by such a pattern. But does that mean that Wittgenstein's interlocutor is right in claiming that there is a particular process of reading that can be recognised by paying attention to what is going on in one's mind?

To this question Wittgenstein responds in more than one way. On the one hand, there is the simple strategy of asking his interlocutor to look and see and to report what he does perceive. On the other hand, Wittgenstein attempts to go along with certain suggestions made by his interlocutor and to elaborate on these suggestions. The first strategy typically leads to fairly straightforward defeat: the interlocutor simply fails to discover what he originally believed to be present. The second strategy leads to new questions and in some cases indicates answers that seem to lie outside the perspective from which the original question had been asked.

Wittgenstein does not find it difficult to admit that there are characteristic sensations accompanying various types of acts of reading. He mentions 'feelings of hesitating, of looking closer, of misreading, of words following on one another in a more or less familiar fashion, and so on' (PI §159). And he contrasts these feelings with those typical of cases in which one recites something one has learned by heart and cases where one pretends to be reading. But this admission does not amount to agreeing with his interlocutor's claim that there is a specific type of feeling that would allow us to recognise an act of reading from within, as it were. It does not even amount to agreeing that normal acts of reading need to be accompanied by any such feelings.

Wittgenstein's interlocutor remarks that in reading aloud the words uttered 'come in a special way'. To this, Wittgenstein replies without hostility but with a request to spell out what his interlocutor means by that. And he explicitly wonders whether the whole idea is not 'a fiction' (PI §166). Under pressure, the interlocutor does not know what to say in answer to the question *how* a certain sound came. He has to admit that he has 'no idea what to say about it'. And asked to describe the way in which the writing down of the letter 'a' comes and whether this way differs from that in which the reading out loud of the same letter comes, he cannot help confessing 'All I know is, I looked at the printed letter and wrote the cursive letter.'

In discussing cases of letting oneself be guided by certain signs Wittgenstein asks his interlocutor to make an arbitrary doodle on a sheet of paper, to copy the doodle and let himself be guided by it (PI §175). When asked to describe what went on, the interlocutor has to concede that he is at a loss. He replies: 'Sure enough, I let myself be guided here. But what was characteristic in what happened? – If I say what happened, it no longer seems to me to be characteristic.'

These are cases of straightforward defeat: the interlocutor fails to discover what he expected to find. But it is typical of Wittgenstein's style of writing in the *Investigations* that it is not always clear whether an identifiable interlocutor or Wittgenstein himself is speaking – perhaps a Wittgenstein in a special mood or conscious of a particular kind of temptation. The continuation of the passage just quoted appears to be spoken in Wittgenstein's own voice, but at the same time there is no obvious break between the interlocutor's words and the following observation which mentions a new aspect of the matter. Whoever the speaker may be, he says:

> ...*while* I am being guided everything is quite simple, I notice nothing *special*; but afterwards, when I ask myself what it was that happened, it seems to have been something indescribable. *Afterwards* no description satisfies me. It's as if I couldn't believe that I merely looked, made such-and-such a face, and drew a line. (PI §175)

We remember that Wittgenstein wondered whether talking about a special way in which words come in the act of reading was not a fiction. The words just quoted provide this suspicion with a special background: it could be that looking back on certain acts makes it appear inevitable to feel that these acts had a particular inner quality – a quality, however, which we cannot describe when we try to give an account of it. So we would be dealing with a kind of projection. But if we are to understand what it is that is projected in such cases, it will be helpful to look at the second strategy I distinguished above: the strategy of going along with all kinds of suggestions about characteristic or special feelings or experiences and see whether they contain anything helpful.

Let us look at some of the cases described by Wittgenstein where he takes up the idea of special feelings or experiences and discusses the question to what extent this idea can be clarified.

The first notion I want to mention in this context is the notion of familiarity. You remember the unfamiliar sign reproduced in §166, which looks a bit like a reversed Sigma. If you encounter it on a printed page, you will react differently to it from the way in which you react to familiar signs. You cannot read it in the same way in which you read the letters of our alphabet. On the other hand, you may associate arbitrary sounds with it or wonder which sound will occur to you if you gaze at it. Or you can say that the sign reminds you of a certain sound – which is something you cannot plausibly say in a case where you are confronted with a letter belonging to our familiar alphabet.

This idea of familiarity is one which Wittgenstein evidently finds absolutely intelligible. He applies it to the way a printed page appears to a reader and writes:

> ... the mere look of a printed line is itself extremely characteristic – it presents, that is, a quite special appearance, the letters all roughly the same size, akin in shape too, and always recurring; most of the words constantly repeated and immensely familiar to us, like well-known faces. – Think of the uneasiness we feel when the spelling of a word is changed. (And of the still deeper feelings that questions about the spelling of words have aroused.) (PI §167)[4]

An important feature of these examples of familiarity is that these are cases in which it is appropriate or natural or correct to *say* certain things. True, we are dealing with *feelings* of familiarity. But there are things that will be said, things that may be said, and things that should not or cannot be said. Of an unfamiliar sign, but not of a familiar letter of the alphabet, can I say that it reminds me of a certain sound? And if an unfamiliar spelling of a word appeals to me or appals me, I shall have to be able to say something reasonable in reply to the question what it is that I find appealing or appalling. In short, what one says about such feelings of familiarity and certain of their consequences is embedded in a context of conceptual connections that find expression in our regarding some phrases appropriate and others unsuitable or incorrect.

An idea which is related to the notion of familiarity is that of forming a unity, which Wittgenstein mentions in a later remark (PI §171). There he says that it might be apt to say that 'when one reads, letter and sound form a *unity* – as it were, an alloy.' He adds that this case is similar to that in which 'the faces of famous men and the sound of their names' are fused together: 'it seems to us that this is the only right one for this face.' This is an example that Wittgenstein mentions more than once; and he likes to illustrate it by referring to the names and pictures of famous composers, in particular Schubert.[5] What is remarkable about this example is that it has clear correctness conditions, as here you may speak of a unity or a case of fusion only if the name concerned is really that of the person depicted. If it is not, you will have a case like that mentioned in a much later remark, where we are asked to imagine someone's attempting to paint a picture entitled 'Goethe writing the ninth symphony'. In such a case one could not imagine anything 'that would not be embarrassing and ridiculous' (PI, II, vi, p. 183).

The second kind of case I want to look at is that of apparent feelings of influence. According to Wittgenstein, this is an idea that may seem plausible when we compare cases of reading letters from our ordinary alphabet with cases of glancing at arbitrary flourishes (PI §169). What I see does tend to have different kinds of impact on me: in the first case it is natural for me to respond in a certain way, for example, by pronouncing the standard sound indicated by the letter. I may even find it impossible to resist this tendency; I cannot help producing or inwardly hearing the corresponding sound. And in such cases it may appear natural to say that one feels an influence – an influence emerging from the letter which I find it impossible to disobey.

In the case of arbitrary marks the matter is different. Here I may *decide* to associate a certain sound with the sign. At a pinch, I may find one sound more suitable than another. But the connection would be extremely tenuous, and I would surely not dream of speaking of an influence which the arbitrary mark works on me.

However, there is something peculiar about this way of talking about feelings of influence. It is a bit of a recherché kind of phrase; it appears contrived, and Wittgenstein brings this out in the following remark:

> We'd never have hit on the idea that we *felt the influence* of the letters on us when reading, if we had not compared the case of letters with that of arbitrary marks. And here we do indeed notice a *difference*. And we interpret it as the difference between influence and absence of influence. (PI §170)

This is a very insightful and very important observation. Two points are made. The first point is that without this sort of comparison we would not have felt inclined to use the expression in question, that is, we would not have spoken of a feeling of influence in the first place. The second point is that the difference we discover when comparing letters and arbitrary marks is *interpreted* by us in a certain way. It is interpreted by means of using expressions that we feel inclined to use only in the context of this comparison. And in this context they are ways of gesturing at something that is not, and perhaps cannot be, fully clear to us. But through *interpreting* the difference in this way the expressions we felt inclined to use in the context of comparison become descriptive phrases which we may then feel justified to detach from this quite specific context. If we so detach them, we shall proceed to speak of feelings of influence in a general way. But that could soon become misleading, as it was only against the background of that comparison that the expression could helpfully be used to convey something.

The third case of special feelings or experiences has already been alluded to. This is the case of experiences of being guided. This time, Wittgenstein explores the matter by starting from a list of examples that are meant to illustrate the notion of being guided. They can be seen as variations on the theme of being guided along a certain path. These examples include cases like the following:

- You are blindfold, and someone leads you across a difficult stretch of countryside
- Against your will someone takes you by the hand and pulls you in a direction where you do not want to go
- You are guided by a partner in a dance
- You walk along a track in a field and let yourself be guided by its course.

As Wittgenstein says, there are similarities between these cases; but as he suggests, there is nothing they all have in common. Some of them involve a certain degree of receptivity: in the blindfold case, you will have to pay attention to the other person's tugs of his hand if you want to avoid stumbling and falling down; in the dancing case you will have to notice and respond to the smallest movement and tension on the part of your partner if you want to enjoy the dance. The case of being led off against your will need not involve any kind of receptivity on your side, and the case of the track in a field is such that following the track may not require more attention than any ordinary case of walking ahead.

Of course, we are meant to apply certain features of these cases to acts of reading, and it does not seem difficult to spot a number of analogies. Receptivity and paying attention are clearly needed if you want to study and understand texts of a certain degree of difficulty, for example, if you want to avoid the pitfalls hidden in a legal document. Skimming the local pages of your newspaper, on the other hand, will often be more similar to simply following a track in a field. Reading the *Philosophical Investigations* can in many ways be compared to being guided by a partner in a dance. And so on. Pointing out such analogies may actually be quite helpful if you are interested in forming an impression of the diversity of cases falling under a given concept. But in view of their not exhibiting one common feature it will not assist Wittgenstein's interlocutor. He repeats his earlier motto and exclaims 'But being guided is surely a particular experience!' (PI §173). Previous to this he had insisted that 'reading is a quite particular process' (PI §165), and this and similar phrases are repeated several times in the section on reading.

The importance Wittgenstein attached to this kind of phrase is underlined by the fact that at some stage after the completion of the last typescript of *Philosophical Investigations* he inserted a slip of paper bearing two short notes on the use of the German phrase *'ein ganz bestimmter.'* In most editions of the book these notes are printed as footnotes. They say: 'The grammar of the expression "a quite particular" (atmosphere). / One says "This face has a quite *particular* expression," and perhaps looks for words to characterise it.'

One thing these notes serve to bring out is a certain ambiguity which one should take note of and bear in mind. The ambiguity is simply this: that the German phrase *'ein ganz bestimmter'* etc.[6] can be read as emphasising, first, that we are talking about one single, individual, case which is unmistakeable in its peculiarity and, second, that we are talking about a certain identifiable *kind* of case. There are different sorts of example that can be used to illustrate the second type of emphasis in contradistinction to the first type. One sort would be a report on a pain I felt: *ein ganz bestimmter Schmerz* could be an occurrence which I find very special, even unique, but which I have never felt before and cannot easily classify; on the other hand, it could be an occurrence of a kind which I am quite familiar with: for example, it's the same sort of pain I had yesterday evening but quite different from the sort I had this morning. Another sort would be descriptions of certain impressions, sensations, feelings. In such cases, my use of an expression like *'ein bestimmter Geschmack', 'ein bestimmtes Aroma', 'eine bestimmte Atmosphäre'* may again mean that I want to underline the individuality or peculiarity of the given case, but it may also mean that I am thinking of a certain *type* of case, such as 'a taste of mint' or 'the aroma of coffee' or 'a desolate atmosphere'.

This ambiguity is marvellously exploited by Wittgenstein in his short reply to his interlocutor's exclamation 'But being guided is surely a particular experience!.' This reply runs as follows: 'The answer to this is: you are now *thinking* of a particular experience of being guided' (PI §173). What the interlocutor has in mind is, of course, a specific *kind* of experience. And what he wants is a kind of experience that is independent of external features like mint or coffee or publicly accessible situations of desolation. Even pain won't do for him, as kinds of pain can always be described in intersubjectively intelligible ways that bring in perfectly public features. For the same reason, the analogies with various kinds of experiences of being guided were no help to Wittgenstein's interlocutor. Their function as analogies depends on the possibility of describing publicly accessible – and publicly decidable – cases that can be found more or less appropriate. And in judging them to be more or less appropriate we apply various kinds of criteria that can be spelled out and whose use can be taught and learned in public.

What the interlocutor has in mind is a kind of experience or feeling *sui generis*, something whose presence I could detect by simply and exclusively paying attention to details of my stream of consciousness. That this is the interlocutor's project is underlined by a short remark added a number of years after the German revision of the *Brown Book* version of the section on reading. This parenthetical remark says: 'Compare with this the idea that memory images are distinguished from other mental images by some special characteristic.' (PI §166b) The idea Wittgenstein alludes to is the familiar one according to which memory images have an internal feature that makes them recognisable as such. (Wittgenstein compares this idea to the difference between faded old photographs and fresh new ones. Naturally, the comparison does not work – and that is meant to discredit the idea of 'faded' memory images.[7])

What Wittgenstein points out to his interlocutor (and to us) is that there is a tendency to believe that the indescribable individuality of the single case can somehow carry over to the general case. This is an illusion. One thinks of – that is, one remembers – a particular, individual occurrence and gains the impression that what appears to be the indescribable qualitative nature of that occurrence is a characteristic of a certain *kind* of occurrence or process. According to Wittgenstein's diagnosis, two aspects of the situation play a particularly important role in engendering this sort of illusion: memory has already been mentioned; the second aspect is linguistic, an almost magical invocation of certain words. Here I shall repeat a previous quotation and add the rest of the paragraph it was taken from:

> ... *while* I let myself be guided everything is quite simple, I notice nothing *special*; but afterwards, when I ask myself what it was that happened, it seems to have been something indescribable. *Afterwards* no description satisfies me. It's as if I couldn't believe that I merely looked, made such-and-such a face, and drew a line. – But do I *remember* anything else? No; and yet I feel as if there must have been something else; in particular when I say *'guidance'*, *'influence'*, and other such words to myself. 'For surely', I tell myself, 'I was being *guided*.' – Only then does the idea of that ethereal, intangible influence arise. (§175)

I think the message is clear. But his success in getting it across is perhaps not sufficient to explain Wittgenstein's satisfaction with the section on reading. The other important feature of this discussion is the sensitivity with which he plays along with all kinds of temptations and tendencies that make themselves felt in thinking about what we are aware of when we do certain things. As Wittgenstein's discussion shows, we manage to get across a good deal of what we feel. In getting it across we often use expressions that suggest themselves to us with particular force. On the other hand, many of the expressions that suggest themselves to us can be misleading; they can be misleading because we easily persuade ourselves that they fit essential inner features of the experience. Here we tend to forget, as Wittgenstein points out, that we do not really 'remember such an experience'. We talk as we do when we speak of feelings of influence or experiences of being guided because when we reflect on what we experience in such a case we 'look at it through the medium of the concept' in question.[8]

Notes

1. This terminology (*Urfassung, Frühfassung,* etc.) is based on the editors' notes in the unpublished but circulated 'Helsinki' edition of *Philosophical Investigations* and was used in my historical-genetical edition of the text (PU). The *Urfassung* is the earliest version (= MS 142 in von Wright's catalogue of Wittgenstein's papers). See von Wright (1982).
2. Wittgenstein's *Brown Book* was edited by Rush Rhees and published in (BB). Quotations from *Philosophical Investigations* are taken from (PI). Occasionally Miss Anscombe's translation has been modified. The German revision of the *Brown Book* was edited by Rush Rhees and published under the title *Eine philosophische Betrachtung* (EPB).
3. The main reason for this may be the fact that my reading of the remarks on rule-following is non-standard, that is, it is neither Kripkean nor anti-Kripkean. See my Schulte (2007). (That Kripke's suggestions are *exegetically* erroneous was pointed out by me as early as in Schulte (1982).)
4. The effects of changes in the standard spelling of words are discussed in several passages in Wittgenstein's writings, for example, MS 137, p. 44b (in the context of meaning-blindness, cf. RPP II, §572). A particularly interesting passage can be found on the first page of MS 150 ('C 6'). Here Wittgenstein writes: 'When wondering what the impression made by a word consists in, the last thing that comes to mind is what we *see* when looking at the word. We suppose that the look of the word (*Wortbild*, cf. PI §§167, 169, 170 [vs. 291]) doesn't really matter and that the impression is somehow situated behind the look of the word (*Wortbild*). And this is a mistake that we make again and again. But the shape of a word – what we have seen countless times, as we have all words of ordinary language – makes a deep impression. Just think of the difficulties we feel when the rules of spelling are changed. Such changes have been regarded as sacrilege...' (my translation). See also MS 152, p. 7. Cf. the chapter on meaning-blindness in Schulte (1993, in particular pp. 68–9).
5. Cf. EPB, p. 184; MS 137, p. 83b; MS 138, p. 12a (=MS 144, p. 78 = PI, II, xi, p. 215 f); MS 169, p. 4v (= LW II, p. 4b [transcription and translation faulty]).
6. See PI §167: 'Now what is there in the claim that reading is "a quite particular process"? It presumably means that whenever we read, *one* particular process takes place, which we recognise – '
7. Cf. MS 135, p. 143; MS 136, p. 4a; MS 138, p. 16b; MS 144, p. 24 (=PI, II, viii, p. 185 e).
8. Cf. PI §177. Some of the relevant connections between concepts and ways of looking at or dealing with the world are discussed in my Schulte (2009).

3
Reasons and Causes

Frederick Stoutland

There has been a sea change in philosophy of action since its revival as a distinct subject some 40 years ago by Wittgenstein-inspired philosophers like Ryle, Anscombe, Melden, Kenny, or von Wright. Wittgensteinian orthodoxy has been replaced by a new orthodoxy – the so-called 'standard story,' which construes action as bodily movements caused and rationalized by inner states or events.[1] This sea change is usually characterized as a shift from the view that reasons for action cannot be causes of action to the view that reasons must be causes, a shift largely credited to Donald Davidson's influential papers.

While admitting the sea change, I think it is too simple, indeed highly misleading, to characterize it in those terms. The relation of reasons and causes is a complex topic that bears on significant philosophical issues that go far beyond any slogans about their relation. In this paper I want to deal with some of these issues on their own merits, consider whether or how Wittgensteinians were wrong about the relation of reasons and causes, and conclude with some remarks on the nature of the sea change.

One philosopher who knew the complexity of the reasons-causes issue is Davidson, and my discussion will center around Davidsonian themes. Although his writings were crucial to the standard story becoming an orthodoxy, he did not accept many of its central tenets. His account was deeply influenced by Wittgensteinian philosophers, above all by Elizabeth Anscombe (he called her *Intention* 'the most important treatment of action since Aristotle'[2]), whose view he accepted in large measure even while arguing that action explanation was causal. His immensely influential work, that is to say, gave rise to an account of action that was not his own.

1

There is a clear conceptual distinction between a reason taken as a consideration that favors an agent's acting intentionally in a certain way, and a cause taken as a relation obtaining between events. This is not the only concept of reason: there is a reason for a car not starting (its battery is dead) or for feeling nauseous (you thought you were on a shaky precipice), but those are not reasons for acting intentionally, and they do not favor the occurrence of the effect. The relevant concept of reason is *normative* in the broad sense of showing the agent's acting intentionally to be good in some way: to be beneficial, appropriate, required by morality or her profession, etc. Agents do not, of course, always act intentionally in a way that would *actually* be good but they do, in general, *think* their acting would be good in some way: they have their own reasons for acting as they do.

A reason for action is *intensional* in that '...is a reason for...' connects statements that express (or are) facts; it is not a relation between singular terms that refer to events or other entities. We might speak of an event as a reason for an action but it is a reason only under a certain description or *as* having a certain predicate. The same is true of what something is a reason *for.* Actions, for instance, are events, but a reason for an action is a reason only for the action under a certain description – *as* such and such. The truth of statements of the form, '*R* is a reason for *A*,' depends, therefore, not on what events or objects are referred to but on the way they are (truly) described.

Causes, on the contrary, are neither normative nor intensional, 'cause' being taken in its primary sense as designating a relation between particular events. That *C* causes *E* is independent of the normative status of *C* or *E* and of whether it would be good in any sense for *C* to cause *E*. If *E* is good in some sense, it might be good that *C* causes it, but that has nothing to do with whether *C* does cause *E*.

Causes are *extensional*: '...is a cause of...' is a relational predicate that relates singular terms that refer to particular events that are causally related. Statements of the form '*C* is the cause of *E*' remain true, therefore, under any substitution of terms for '*C*' and '*E*' that refer to or describe the same events as *C* and *E* respectively. How the events are described is irrelevant to their being causes or effects because nothing is a cause or an effect under one description and not another, or *qua* having one predicate and not another. It is particular events as such that stand in causal relations.

Many philosophers have generalized the causal relation and asserted that if one event is the cause of another, all events of the same kind as the cause are causes of events that are of the same kind as the effect. This might be true, but it would be false to claim (with Hume) that the assertion that one particular event *C* caused another *E* *meant* that all events of kind *C* are always followed by events of kind *E*. That claim conflicts with our ability to *observe* one event causing another, which would not be possible if causal claims were inherently universal. Some philosophers have argued that what *explains* the causal relation between particular events is that they are instances of a general law, but even if that were true, it would not yield a substitute for that relation between particular events. In my view, it is not true: events are not causally related *because* of a causal law but, on the contrary, it is instances of singular causation that ground causal laws.

Anscombe wrote this about the primacy of observable causes over a general concept of causation:

> ...In learning to speak we learned the linguistic representation and application of a host of causal concepts. Very many of them were represented by transitive and other verbs of action used in reporting what is observed.... The word 'cause' can be *added* to a language in which are already represented many causal concepts. A small selection: *scrape, push, wet, carry, eat, burn, knock over, keep off, squash, make* (e.g., noises, paper boats), *hurt*. But if we care to imagine languages in which no special causal concepts are represented, then no description of the use of the word in such languages will be able to present it as meaning *cause*.[3]

Strawson put Anscombe's point about the diversity of causes by writing that there are 'specific varieties of the bringing about of effects by things' – different 'types of effect-producing actions.' But this does not entail that an event causes an effect only under a description – in virtue of some property. An event causing a certain effect is a natural relation between the events that varies with the type of cause and effect involved. But that is not like distinguishing between the causally relevant (and irrelevant) properties of the events. Events that are instances of a certain type of causal relation (or the manifestation of a certain kind of causal power) are so under *every* description. The fact that causal relations are of different types and involve different types of changes in different types of objects does not entail that events are causally related *because* they are instances of one of those types or of a general law relating them. None of these points undermines the extensionality of the causal relation.

This is also Davidson's view even though he defended the nomological character of causation: if events are causally related, there must be a strict law instantiated by true descriptions of those events. We need not know those descriptions or the law, but since laws are strict only if the events described belong to a closed system (are such that whatever can affect the system is part of that system), and since only physics describes a closed system, we know they are laws of physics. This is consistent with causes being extensional since Davidson denied that events are causally related because of any strict laws: 'It makes no literal sense to speak of events as causing things because of, or in virtue of, anything.'[4] On his view, that there is a causal relation between events entails that there is a strict law connecting physical descriptions of the events, which does not mean that the events are connected in virtue of such a law.

2

The claim that reasons are not causes is surely acceptable on this understanding of the concepts, which is how Wittgensteinian philosophers of the 50s and 60s typically understood them. Anscombe's argument, for instance, that reasons for action are not what she called 'mental causes' ('what went on in my mind and issued in the action') turned largely on the notion that mental causes are extensional and reasons are not. Others made similar arguments, even if sometimes rather carelessly, which appealed to the conceptual distinction between reasons and causes I have articulated.

The latter does not, of course, entail that the same items cannot in some sense be both reasons and causes. What it does entail is that, although an event as such can be a cause, it can be a reason only under a description: reasons, therefore, are not events but statements or facts that involve events. Events as such can be causes but not reasons, and reasons cannot be events but conceptually structured items that involve events.

This is particularly important given the centrality of the concept of intention to an adequate account of action. I agree with most philosophers that actions are events, but I also agree with Anscombe and Davidson that no event is an action unless it has a description under which it is intentional. The concept of intentional is *intensional*, and hence an action *as such* is not intentional: it is intentional under some descriptions and not intentional under others. This is crucial given the Anscombe-Davidson account of the individuation of action: an agent who (to paraphrase Anscombe's example) (1) moves his arms up and down, thereby (2) operating the pump, (3) wearing out his gloves, (4) bumping the man standing near the pump, (5) filling the water supply with poisoned water, (6) thereby killing the occupants of the house, is acting only once in moving his arms up and down, the rest following from that without further acting on his part. His acting has, however, many descriptions in terms of what results from the movements of his arms. His acting is intentional under descriptions 1, 2, 5, and 6 (provided his moving his arms has the necessary results) and not intentional under descriptions 3 and 4 (it is inadvertent under 3, an accident under 4).

Being intentional is not, therefore, a *feature* (or property) of an action: to say of an action that it is intentional is not to ascribe a description or predicate to it. If it were, an action would have inconsistent features, being both intentional and not intentional. If an action is intentional under description *D*, then *'D' is* a predicate that designates a feature of the action, but 'being intentional' is not. An action is intentional only under a description, which does not mean there are such things as 'actions under a description'; there are only actions, which are intentional under some descriptions and not intentional under others.

The claim that an event is an action only if it is intentional under some description is not dispensable: an action is *essentially* intentional and hence *must* be intentional under some description. At the same time, being intentional is not a feature of an action. To understand how both these points are true, note that, although a species of the genus *action*, *intentional* action is not a species in virtue of having a feature that distinguishes it from other species of action or from the genus. The relation of intentional action to action is not like the relation of square to four-sided figure but like the relation of pure gold to gold. The genus is *gold*, and the species are *pure gold* and *impure gold*, neither of which is a species in virtue of having special features that distinguish them from the genus. Both pure gold and impure gold are gold, but pure gold is *essentially* what gold is, while impure gold is a lesser species that exhibits only imperfectly what gold really is

Another example, one closer to action, takes the *human body* as a genus, of which a *healthy body*, a *diseased body*, and a *dead body* are species. What distinguishes the species from the genus is not some feature each has that the genus does not: adding a feature to a human body does not yield a healthy, diseased, or dead body. Rather, the healthy body is *essentially* a human body, while a diseased body is a degenerate form, and a dead body is a human body only in shape and origin. What a human body *is* is exhibited by a healthy body, the other species being human bodies only in that they resemble, once resembled, or fall away from a healthy body.

The application to action takes the genus to be *actions*, the species to be *intentional* action and *non-intentional* action. Intentional action is an *essential* species of action, not a species in virtue of having a distinguishing feature (being intentional cannot be a feature of action) but what action is essentially, which exhibits in an undistorted way what it is to act. We act non-intentionally, therefore, only *in* acting intentionally: acting inadvertently or by mistake are lesser forms of acting, while the by-products, as it were, of acting – wearing out our shoes, disturbing the molecules in the air, causing electrical impulses in our brain – are acting only in a peripheral sense.

A similar point holds for bodily movements (or movings).[5] The genus is *bodily movements*, the species are *intentional* bodily movements and *non-intentional* bodily movements. Intentional bodily movements are the essential species. They are the normal movements of limbs by agents capable of moving them in a normal way, where *limbs* are identified precisely as what agents can move directly and intentionally. Non-intentional bodily movements are of two kinds. There are, first, the movement of limbs that are crippled or diseased in some way: we move our limbs in less than intentional ways while moving, or trying to move them, intentionally. There are, second, *mere* bodily movements – 'random movements' – that result from a tic, a brain probe, electric shock, fright, etc. They are only accidentally the movements of limbs for they may be movements of parts of the body that are not limbs.[6] Since action is essentially intentional, whatever accounts for an event being an action must account for its being intentional under a description but not account for its having any distinctive feature. We do not add a feature to an event to get an action or add a feature to an action to get an intentional action. Nor is there a feature we can subtract from an action to get a mere event, or a feature we can subtract from an agent's intentionally raising her arm to get her arm's going up. The quest for such a feature to add or subtract is a confused and hence futile quest.[7] It follows that no *cause* of an event can make it the case that the event is an action and hence intentional under a description. A cause brings about an event with all its features and hence all its descriptions and does not differentiate the descriptions under which it is intentional from those under which it is not. The reason, of course, is that causation is extensional: *C* causes *E* under every description of *E*; that is, it brings about *E* with all its features.

Philosophers who attempt to account for actions being intentional in terms of their cause confront the problem of *deviant causation*. The problem arises because if a cause can bring about an action that is intentional, the same cause can bring about one that is not intentional (or not even an action). Philosophers who confront the problem attempt to resolve it by adding conditions to the cause of an action in order to differentiate the cause of an intentional action from the cause of a non-intentional one. An action is intentional, it is argued, only if it is caused 'in the right way,' that is, under conditions that differentiate its cause from the cause of a non-intentional action. But this is hopeless if causation is taken as extensional: no matter how the action is caused, no matter the conditions under which it is caused, the action is caused with all its features, but it will be intentional under some of them and not intentional under others.

3

The obvious move at this point is to bring in the notion of *explanation* and the explanatory role of both reasons and causes. Explanation is an intensional notion: '...explains...' does not connect terms that refer to events or other entities but statements that are (or express) facts. It expresses a conceptual connection, and hence events or other entities have an explanatory role, whether as explaining or being explained, only under some descriptions and not under others. A statement of the form '*A* explains *B*' does not remain true under any substitution of terms that refer to *A* or *B*; whether it is true depends on how *A* and *B* are described.

To construe reasons for action as I have, as considerations that favor an agent's acting intentionally in a certain way, is to construe them in their normative role. But reasons also have an explanatory role: they not only favor an agent's acting but can explain why she acted intentionally in that way. To answer the question of what constitutes events as intentional if not their (extensional) causes, we must consider the explanatory role of reasons. Since our concern is with bodily (not mental) acts, our question is what constitutes bodily movements as actions, hence as having descriptions under which they are intentional.

Let me note first another rationale for why, as Anscombe put it, 'We do not add anything attaching to the action at the time it is done by describing it as intentional.'[8] On the one hand, we do not add a feature to an *action* by describing it as intentional because, as action, it is already intentional under a description. On the other hand, to add a feature to bodily movements that are not actions in order to constitute them as intentional would be to add a feature to *non-intentional* movements; but the latter are the movements either of diseased limbs or random movements of various parts of our body. Nothing we add to *them* could possibly constitute them as intentional, to say nothing of constituting them as *essentially* intentional under a description.

The answer to what constitutes bodily movements as intentional is that they have a certain kind of explanation, namely, one in terms of agents' reasons for acting as they do. An agent who acts because of his reason for so acting *necessarily* acts intentionally (his reason cannot explain his acting inadvertently, accidentally, or in ignorance), but being explained by his reason is not a *feature* of his action because it explains his action under some descriptions (those under which it is intentional) and not under others. Since being explained by an agent's reason is not a feature of an action, it does not constitute an action as intentional by adding a feature to it.[9] Consider Anscombe's man who pumps. He moves his arms up and down, thereby pumping and filling the water supply, his intention being to poison the inhabitants of the house. Accomplishing the latter was his reason for moving his arms so as to operate the pump to fill the water supply, and hence his movements were intentional under the descriptions 'operating the pump', 'filling the water supply', and 'poisoning the inhabitants' (assuming he accomplished those things). He also wore out his gloves (which he may have known he was doing), but he had no reason to do so; if he had acted for that reason, it would not have been inadvertent. The contrapositive is also true: because wearing out his gloves is not a description under which his bodily movements were intentional, it was not his reason for those movements. Acting for a reason and acting intentionally are conceptually inseparable.

There is a circle here that is inescapable. On the one hand, an agent's reason for acting explains why he is acting intentionally under a description, not why he is acting under descriptions of his action whereby it is inadvertent, accidental, mistaken, and so on. On the other hand, acting is (in general) intentional under a description only if the agent acts for a reason conceptually linked to that reason. I say 'in general' because there are exceptions: one may, for instance, whistle or hum a tune intentionally but for no reason. Such an action is intentional, however, only if the agent could have performed that particular action for a reason. These exceptions are uncommon because an agent's acting intentionally for no reason is intelligible only in the context of his generally acting for a reason and knowing he could have had a reason for what he did intentionally.

'Reason' here means *constitutive* reason, the reason that constitutes an act as intentional. While constitutive reasons are crucial to a philosophical account of action, they are usually not very interesting to empirical researchers about reasons for action, who claim, for instance, that the reason people buy a certain car or vote for a certain candidate is often obscure or unknown to them. They maintain that the reason people think they have for buying a car or voting for a candidate often does not explain *why* they bought that car or voted for that candidate. What is claimed in such contexts to be the reason for which an agent acted is often, however, not that at all but rather an explanation of why she has the intention with which she acted, where the intention is the constitutive reason. When it is claimed that the agent is, indeed, mistaken about her own reason (and not about the explanation for *having* the reason), it is her longer term reason that is in question, and that is not the reason that *constitutes* the action's intentionality.

For example, the man who pumps may think that the reason he is pumping is to poison the inhabitants of the house who are brutal dictators. But he may be mistaken: his reason for pumping may be simply to collect the huge sum of money promised him. It may, then, be unclear whether he is *intentionally* poisoning the inhabitants of the house, but that is irrelevant to whether his *pumping* is intentional, since his poisoning the inhabitants is not (in this case) a *constitutive* reason. The constitutive reason for his moving his arms up and down may be as short term as his thereby intending to operate the pump or perhaps to fill the water supply. The latter do not typically interest psychologists or sociologists of human behavior, but they are decisive for an account of what it is for an action to be intentional.

4

Constitutive reasons are decisive because to act intentionally is to know what you are doing in a sense that requires knowing the constitutive reasons for your acting. This is the commonsense view, which I shall make explicit and defend.[10]

Acting is intentional under a description just in case the agent knows in a particular way what he is doing under that description. His knowledge cannot be a matter of observation, because one can observe one's bodily movements even if they are not intentional, and because one typically knows what one is doing intentionally without observing either one's bodily movements or their results. Nor can the knowledge be introspective since we can, at best, know only our own mental states that way. To act intentionally, we must know more than our own intentions and other mental states and, normally, more than how our limbs are moving (we need, in fact, not always know that); we must know what happens as a result of moving our limbs. Only if the man who pumped and replenished the water supply knew what he was doing under those descriptions – knew the pump moved and the water supply got filled up – was he doing so intentionally. Moreover, only if he was doing so intentionally could he know without observation what he was doing.

If such knowledge of one's intentional actions were *theoretical*, it would be totally mysterious. It would, for example, be a mystery how a man could knowingly be filling the water supply when he cannot observe and does not infer that he is doing so. Theoretical knowledge is derived from the object known, for instance, knowing that the water supply is being filled by observing the water supply or making inferences about it. If one makes a mistake on that basis, then one has made a mistake in one's judgment. *Practical* knowledge, on the contrary, is knowledge of what we are doing that is derived not from observation or inference about what is happening but from our intentionally doing it. The man knows he is replenishing the water supply because that is what he intends in moving his arms up and down and operating the pump. He knows what is happening because he is intentionally doing what happens. He may, of course, be mistaken – the water pipe might, unknown to him, have broken – but his mistake is not that he has insufficient evidence for what is happening but that he fails to do what he intended. His mistake is not in his judgment but in his performance. He was wrong *about* what he did only because he was wrong *in* what he did.

A man may throw a rock over a high wall intending to hit his enemy on the other side, and if he hit him, he will have done so intentionally even if he never knows whether he actually did – never knows what happened on the other side. But the knowledge he may never have is theoretical not practical. If he did hit him, he may never know it in the theoretical sense but he does in the practical sense because he did what he intended to do: there was no mistake in his performance. Of course, if he missed him, he knows it neither in the theoretical nor in the practical sense.

What an agent *is* doing intentionally under a description must be what he *knows* he is doing intentionally, which means he knows the intention with which he is doing it, which is the constitutive reason for his acting. As Anscombe wrote, 'It is the agent's knowledge of what he is doing that gives the description under which what is going on is the execution of an intention.'[11] If the man is intentionally replenishing the water supply, he must know he is so doing, and if he does not know that, then no one else can know it either, for practical knowledge of what one is doing, hence of the intention with which one is doing it, is constitutive of one's doing it intentionally.[12] If the man intended but failed to fill the water supply, then he failed in his performance and whatever he did intentionally, it was not *that*. What he did intentionally was something that satisfied some lesser intention, perhaps merely moving his arms up and down, and of that he did have practical knowledge.

5

What constitutes an agent's bodily movements as intentional under a description is their being explained by the agent's reasons for acting under that description. I have focused on the intentions with which an agent acts since I take their explanatory role to be primary in constituting actions as intentional. I have noted a circularity here – events are constituted as intentional actions because they are explained by an agent's reasons, but the latter explain events only if they are intentional under a description – which my discussion of practical knowledge was meant to show as harmless. What we have is a wide-ranging network of interconnected concepts: actions, intentions in acting, intentions to act, practical knowledge of what one is doing, reasons for acting, explanations of action in terms of those reasons, and so on. These concepts can be elucidated only together, and none is fundamental to the others.

An agent's reasons for acting intentionally are much more extensive than her intentions. Anscombe, for instance, listed intentions as only one of three sub-classes of what she called 'motives,' while the majority of philosophers take an agent's beliefs and desires to be her reasons for acting. I take these further reasons to have a role primarily in explaining why the agent has the intentions with which she acts, and hence to explain actions by way of explaining intentions. But specifying the precise nature of reasons and their relation to intentions is not crucial for our central question about reasons and causes.

Our discussion has transformed that question into a better form, for it is now about *explanation*: whether an explanation of an agent's intentional action in terms of her reasons – a rational explanation – is a *causal* explanation. Agents may have reasons to act but fail to act on them, or they may have diverse reasons for acting in a certain way but only act on one, the others being explanatorily idle. The one they act on *explains* their acting intentionally as they did, the question being whether such explanation is *causal*, and if so, in what sense.

I will consider this in relation to Davidson's work because he made the question central to philosophy of action and because he took his answer to it to be the point on which he disagreed with the Wittgensteinians whose work he otherwise admired and learned from. He would not have rejected most of what I have so far claimed in this paper, although no doubt formulating it differently. But he surely took himself to differ fundamentally with Wittgensteinian philosophers on the question of whether rational explanations are causal, and I want to consider how fundamental that difference is.

Doing so calls for determining what Davidson's real view is, which requires distinguishing two interpretations of his account of rational explanation and related matters. I call the first the 'standard' interpretation because it is so often cited as a source for the standard story about action.[13] I shall argue that it is a mistaken interpretation that yields an indefensible view and then turn to what I consider a better way to read Davidson.

6

The 'standard story' holds that intentional actions are those bodily movements caused and rationalized by an agent's beliefs and desires that are conceived of as inner states or events in the mind/brain. The story is typically embedded in a physicalist ontology, which Davidson is said to have promoted because, as Jaegwon Kim put it, his 'main task has been that of finding for mind a place in an essentially physical world... [in which] we find nothing but bits of matter and increasingly complex aggregates made up of bits of matter.'[14] While Kim thinks Davidson failed in this task in not being reductive enough, he credits him with an influential attempt to construct a physicalist and causalist account of action.

On this interpretation, Davidson's account is as follows. Reasons for agents to act intentionally are her states of belief and desire, which explain her acting only if they cause her bodily movements. Beliefs and desires as such are inner states, not events, and hence not causes, but they can easily be conceived as events if we think in terms of their changing or coming to be. If they cause bodily movements in the right way (not, for instance, by way of causing nausea or excitement), then the movements are not mere happenings but actions done for the agents' reasons.

The traditional view of this causation was the covering law model, which holds that beliefs and desires cause bodily movements *in virtue of* causal laws that connect kinds of beliefs and desires with kinds of bodily movements, where the kinds are intentional (or psychological) kinds that account both for the causal power of the beliefs and desires and for their being reasons for agents to act.[15] Davidson argued that there are no such causal laws and hence that reasons do not cause bodily movements in virtue of causal laws at the intentional level. But on this interpretation, he held that reasons cause bodily movements in virtue of *some* causal laws, which are at the sub-intentional level. He worked this out by claiming that events are related as causes and effects because all events that are in causal relations are *identical with* physical events that have a description that is an instance of a causal law of physics. Beliefs and desires are causes of bodily movements, therefore, because the beliefs, desires, and bodily movements are physicalistic events that instantiate a law of physics.

As an account of action, this has serious difficulties, which defenders of the standard story have tried diligently to resolve. The most discussed difficulty is that it is an *epiphenomenalist* account in holding that beliefs and desires cause bodily movements not *as* reasons but *as* physicalistic events instantiating laws of physics. While beliefs and desires may, on this account, be reasons that rationalize bodily movements, they do not cause them *as* reasons, and hence causality and rationality fall apart in a disastrous way.

This is a devastating criticism of Davidson interpreted in this way, but it should not be dealt with by turning to reductive physicalism or by returning to the covering law account of rational explanation, but by getting a better reading of Davidson's work. I have already noted that his principle of the 'nomological character of causation' is not that events are causally related *in virtue of* causal laws. On his view, events are not causally related in virtue of anything, hence not in virtue of being physical – or in virtue of being mental. To read him otherwise is to blur the distinction he insisted on between causal relations as extensional and causal explanations as intensional.

In claiming that reasons cause bodily movements, Davidson did not mean that there is a causal relation between reasons (beliefs and desires) and bodily movements that constitute the latter as intentional. That beliefs and desires are not events rules that out, but it is ruled out even if beliefs and desires are construed in terms of their changing or coming to be, hence as events. For the latter would yield an account of what constitutes bodily movements as intentional in terms of causal relations, which would be subject to the criticisms I made earlier of attempts to account for intentionality in extensional terms.

When Davidson claimed that reasons cause actions, he meant that rational *explanation* is a kind of *causal* explanation. The point of explanation in general is to render phenomena intelligible, and what does so under one description of the phenomena may not do so under another. If a cause is to *explain* an effect, therefore, both cause and effect must be described in explanatorily relevant ways. If the explanation is to be *causal*, it must, on Davidson's view, meet three conditions:

1. It must explain the occurrence of an *event* (or what *entails* the occurrence). The explanatory cause need not be an event; typically it is a state of some kind. Hence, if '*A* causes *B*' means that *A causally explains B*, then *A* need not be an event.
2. It must entail the occurrence of some event *associated* with the cause that is *causally related* to the effect. We need not know which event that is, though it must have a physical description that is connected by a causal law of physics to a physicalist description of the effect.
3. Its validity must depend on generalizations connecting descriptions of the cause and the effect that are empirical but *rough* and hence not strict laws.

Davidson claimed that a *rational* explanation is causal because it meets all three conditions. It meets the first because bodily movements are events. The *cause* that *explains* them, however, is not (and need not be) an event, because beliefs and desires are not events but *states* ascribed to whole persons (not to their minds or brains).

It meets the second because when an agent acts *because of* a reason, there is, Davidson contends, an event *associated* with the reason that is *causally related* to the action. This event might be coming to have a belief, changing one's mind, noticing the corner ahead, a change in the brain, etc., etc. It does not causally *explain* the action because it is not the reason (that is the belief and desire) and no one need know what event it is, but it is, Davidson insists, a necessary condition for the explanation to be causal. It is part of what it *means* to say an agent acted for a given reason but it does not explain why he acted for that reason rather than another.

It meets the third condition because its validity depends, in Davidson's view, on empirical generalizations that are not strict laws. Generalizations are necessary because of the dispositional character of beliefs and desires, but they are rough and hold *only* for the particular agent whose action is being explained.[16] If we try to extend them to agents more generally (by adding *ceteris paribus* clauses), they cease to be empirical and become conceptual or normative principles that we use in interpreting agents and their actions if we must verify what they are doing and why.

This reading of Davidson differs from the standard reading in a number of ways. First, there is no claim that beliefs and desires are physicalistic because Davidson's argument that events have physicalist descriptions does not apply to states. 'Beliefs, desires, and intentions,' he wrote, 'belong to no ontology.... When we ascribe attitudes, we are using the mental vocabulary to describe people. Beliefs and intentions are not little entities lodged in the brain.'[17] Second, the events that are causally *related* to an action are not *reasons* for the action; they may have no connection with the reasons and we need not know what they are.[18] Third, these causes do not explain why an agent acts for one reason and not another. Finally, when we must establish what reasons explain another agent's action, we do not appeal to causal generalizations or experimental manipulation, or seek to uncover events that trigger her acting. We rather seek the best overall interpretation of her action in the light of her attitudes, her situation, what she has done and goes on to do, and other considerations that might be relevant.[19]

7

Davidson's claim that an agent acts because of a reason only if her reason causes her bodily movements is much more nuanced than the view typically cited by defenders of the standard story and is, indeed, opposed to it in fundamental ways. Does that mean that Davidson's account of intentional action as bodily movements causally explained by an agent's reasons is not so very different from the best Wittgensteinian accounts? There surely are differences but, taken as a whole, his account is closer to them than it is to the standard story.

Davidson defended his causal view of rational explanation against two objections made by many Wittgensteinians. The first appealed to Hume's thesis that causal explanations require general laws, the objection being that there are no general laws covering reasons and actions, a criticism von Wright, for example, made because he accepted Hume's thesis. Davidson rejected both covering laws and Hume's thesis, claiming that rational explanations are causal in a non-Humean sense. But that is not at odds with Anscombe's view, for she also rejected both covering laws and Hume's thesis. As for von Wright, while agreeing with Davidson that *rational* explanations did not involve covering laws, he thought *causal* explanations required them and hence denied that rational explanations are causal. This is not a fundamental difference because von Wright was happy to say that rational explanations are causal in a non-Humean sense: 'Those who think that actions have causes often use "cause" in a much broader sense than I do when I deny this. Or they may understand "action" differently. It may very well be, then, that "actions" in their sense have "causes" in theirs.'[20]

The second objection (also credited to Hume, who asserted that cause and effect are 'distinct existences') was that conceptual connections exclude causal connections, and hence that the conceptual connections between reasons and actions entail that reasons do not causally explain actions. While Davidson regarded conceptual connections between reasons and actions as crucial to rational explanation, he rejected the objection by appealing to the distinction between causal relations and causal explanations. The claim that cause and effect are distinct existences applies only to *events* and hence only to causal relations between events. *Conceptual* connections hold only between sentences, facts, or descriptions and hence are relevant only to causal *explanations*. The claim that causes and effects cannot be conceptually connected is, therefore, either nonsense or false. It is nonsense to speak of *events* as conceptually connected, while it is false to claim that *descriptions* of events (even if they are causes and effects) cannot be conceptually connected. Whether descriptions are conceptually connected is independent of whether the events described are causally related.

Davidson's refutation of this objection is decisive and surely undermines a number of arguments made by Wittgensteinians. It does not, however, undermine von Wright's distinctive use of the argument (though I think it is faulty in other respects), and it is not one to which Anscombe appealed. His rejection of both these objections to the causal nature of rational explanations does not, therefore, distinguish his view from that of the best of the Wittgensteinians. The main difference between their views on the reason-cause issue is not over the nature of rational explanation but over what it is for a rational explanation to be causal. Two features Davidson regarded as essential for an explanation to be causal are ones they would – rightly in my view – reject as inessential.

The first is that to be causal, an explanation must involve a particular kind of causal *relation*, namely, that an event *associated* with the reason causes the agent's bodily movements. The main reason Davidson took this as essential is (I think) his belief that the difference between an agent merely having a reason and her acting because of it is not in the content of the reason but is additional. Maria's reason to buy a book is that she needs it for her work; if she does not buy it, she merely has the reason, but if she buys it because of her reason, there is an associated event that causes the bodily movements that result in her buying the book. The reason is the same in both cases, but in the second there is in addition to the content an event that causes her bodily movements.

Davidson did not think this account *explains why* an agent acted because of the reason she did because he held that a rational explanation 'provides no reason for saying that one suitable belief-desire pair rather than another (which may also have been present in the agent) did the causing',[21] that is, was associated with an event that caused the action. His aim was to elucidate what we *mean* by the assertion, 'She acted because of the reason', not to give an account of why she acted because of it.[22]

Nor did he hold that to verify that an agent acted because of a reason requires verifying that an associated event caused it. His view that what an agent does and her reason for doing it are conceptually connected means they cannot be verified independently. This sets up an interpretive circle, and there is no appeal except to interpretation in order to verify whether an agent acted *because* of a reason.[23] Having established a plausible interpretation of an agent's reasons and actions, we do not establish *in addition* that there was an associated event that caused her action, since, on Davidson's view, the interpretive conclusion that she acted because of a certain reason *entails* that there was an event associated with that reason.

This meets one objection to Davidson's account but strengthens another since it implies that *establishing* that there is an associated event comes *after* having established an explanation of the agent's action, which means the event entailed by the explanation plays no role in establishing that explanation and hence no role in determining its force. To claim that such an event is always entailed seems unobjectionable simply because 'associated event' is so broad there can hardly fail to be any, so even if there are such events whenever we act, there is no reason to think there *must* be for the explanation to be valid. They play no essential role in a rational explanation.

The second feature to be rejected as inessential involves the *nomological character of causation* – that causal relations entail a causal law at the level of physics. Davidson wrote that 'the ground floor connection of causality with regularity is not made by experience, but is built into the idea of objects whose changes are causally tied to other changes.... Events are as much caught up in this highly general net of concepts as objects.'[24] While that claim may be true, it is obscure what it implies, and it does not establish that whatever regularity is connected with causation entails laws of physics – that, as Davidson puts it, 'singular causal statements entail the existence of strict laws.'[25] Even if it does, it does not justify – and Davidson did not think it is justified – the claim that causation is essentially physical, that it to be grounded in physics, or that good causal explanations depend on good explanations in physics.

If we drop these two features because they are inessential for an explanation to be causal, is rational explanation, as conceived by, say, Anscombe or von Wright, *causal* explanation? I see no reason why not.[26] Explanation in general renders some phenomena intelligible, and what is essential to *causal* explanation is that it renders intelligible why some phenomena came to be or ceased to be. A rational explanation, then, is causal if it renders it intelligible *why* an agent intentionally acted or ceased to act: if it specifies the reason why – because of which – she acted. Wittgensteinians like Anscombe and von Wright (and many others) surely aim to specifying *why* agents acted as they did. They recognized that rational explanation could be characterized as redescribing the action, especially given the view that the same action has many descriptions, but they contended that redescriptions of action are necessary for explaining why agents act. Rational explanations were causal because of the redescription they involved, not in spite of it. We redescribe the man's moving his arms up and down as his operating the pump, filling the water supply, and poisoning the inhabitants. But those can also function as reasons why he moved his arms up and down – as explaining his action – and in so doing they are components of a causal explanation.

It is a different kind of causal explanation than what is given in the natural or behavioral sciences. But Davidson also insisted on a difference between rational and scientific explanation. He wrote that 'there is an irreducible difference between psychological explanations that involve the propositional attitudes and explanation in sciences like physics and physiology',[27] and he endorsed Collingwood's view that 'the methodology of history (or, for that matter, of any of the social sciences that treat individual human behavior) differs markedly from the methodology of the natural sciences.'[28]

8

My conclusion is, then, that the slogans of the 50s and 60s that reasons are not causes should be abandoned, and we should happily agree that reasons are causes, that is, that rational explanations are causal explanations. But my articulation of the latter hardly represents a sea change, for I have argued that the continuities between the best Wittgensteinian accounts and a defensible conception of action more or less like Davidson's are more significant than the differences.

The sea change in philosophy of action is not a shift from the view that reasons cannot be causes to the view that they must be causes. It is rather a shift from a philosophy of action that takes seriously Wittgenstein's rejection of the Cartesian distinction between the physical and the mental to one that is committed to constructing a theory of action in a framework largely determined by physicalist assumptions. The standard story is attractive to many contemporary philosophers because they are able to construe it to fit the physicalism that dominates philosophy of mind. They may not aim to reduce rational explanations to explanations in the physical sciences, but they agree with Hartry Field that 'it is beyond serious doubt...that there is an important sense in which all facts depend on physical facts and all good causal explanations depend on good physical explanations.'[29] This leads to downplaying the normative, to finding a central place for (extensional) causal relations, to construing causal explanations as fundamentally a matter of causal laws, to rejecting any irreducible difference between rational explanations and explanations in the natural sciences.

Some defenders of the standard story are dualists rather than physicalists, but they take for granted the physicalist assumption that philosophy of action begins by asking how mere bodily movements – the bodily movements of physiology and neural science – become actions, in particular what features must be added to them to constitute them as intentional. Physicalists think that the beliefs and desires whose causing is constitutive are states of the brain, while dualists think they are states of the mind, but both think they are inner states rather than states of whole persons, thus missing the Wittgensteinian and Davidsonian critique of this Cartesian view of the attitudes. Their claim that reasons cause actions is the claim that inner states or events cause them, and that *is* a sea change from the conception of rational explanation defended by Wittgensteinians. But the latter can claim that rational explanation is causal explanation, although in a distinctive way, and that I have argued, though not a sea change, is a far better way.

Notes

1. On 'standard story' cf. Smith (2004, p. 196).
2. From the cover of the 2000 Harvard University Press edition of her *Intention*.

3. (Anscombe, 1981).
4. (Davidson, 2005, p. 196).
5. I discuss this issue in Stoutland (2009).
6. This discussion of the essential species and its application to action comes from the writings of Anton Ford of the University of Chicago, particularly his (2008), unpublished PhD dissertation at the University of Pittsburgh, called 'Action and Generality.' It has, of course, clear links to Aristotle.
7. I take it this to be the lesson of Wittgenstein's famous question, '... What is left over if I subtract the fact that my arm goes up from the fact that I raise my arm?' (Wittgenstein, 1953, Part I, § 621). That is a question posed by an interlocutor, not an expression of Wittgenstein's own view, which would reject the question.
8. (Anscombe, 1957, p. 28).
9. This is following Anscombe, who wrote: 'What distinguishes actions which are intentions from those which are not? The answer that I shall suggest is that they are the actions to which a certain sense of the question "Why?" is given application: the sense is of course that in which the answer, if positive, gives a reason for acting.' (Anscombe, 1957, p. 9) Later she wrote: '... The description of something as a human action could not occur prior to the existence of the question "Why?"' (Anscombe, 1957, p. 83).
10. What follows is pretty much what Anscombe argued in her *Intention*.
11. Ibid., p. 87.
12. For an excellent discussion of this point, cf. Moran (2004).
13. It is also often cited by Wittgensteinians who consider Davidson's account of action as fundamentally at odds with Wittgenstein's work. I interpreted Davidson that way for a long time, and as a result a number of the criticism I made of Davidson missed their target. See, for example, my (1985). I came to see the light when I wrote 'Intentionalists and Davidson on Rational Explanation' (Stoutland, 1999, pp. 191–201).
14. (Kim, 2003, p. 113).
15. To put the point in other words, the covering law theory entails that descriptions of beliefs and desires as reasons are also description of them as causally effective, which is to say that the property that makes an event a reason is also the property that makes it a cause.
16. 'The laws implicit in reason explanations are simply the generalizations implied by attributions of dispositions. But then the "laws" are peculiar to individuals at particular moments.' (Davidson, 1980, p. 265).
17. (Davidson, 1999b, p. 654).
18. 'Sometimes the answer [to "Why did you do it?"] will mention a mental event that does not give a reason: "Finally I made up my mind." However, there also seem to be cases of intentional action where we cannot explain at all why we acted when we did. In such cases, explanation in terms of primary reasons parallels the explanation of the collapse of the bridge from a structural defect: we are ignorant of the event or sequence of events that led up to (caused) the collapse, but we are sure that there was such an event or sequence of events' (Davidson, 1980, p. 13).
19. For further discussion and defense of this way of reading Davidson, see Stoutland (1999).
20. (Von Wright, 1971, p. viii).

21. (Davidson, 2004, p. 109).
22. This is contrary to Mele, who offers this as the causal theory's view: 'In virtue of what is it true that he mowed his lawn for this reason and not the other, if not that the reason (or his having it) and not the other, played a suitable causal role in his mowing the lawn.' (Mele, 2003, p. 70).
23. Davidson did not hold that in order to know an agent's reasons and actions we must interpret or verify them. We may, for instance, know such things simply by observing an agent.
24. (Davidson, 1985, p. 227).
25. (Davidson, 2005, p. 219).
26. There are other differences that remain, for example, whether beliefs and desires are the primary reasons for intentional action.
27. (Davidson, 2004, p. 101).
28. (Davidson, 2005, p. 282).
29. (Field, 1992, p. 285).

4
Was Wittgenstein Wrong about Intentionality?

Alberto Voltolini

At least *prima facie*, there is no doubt that the later Wittgenstein conceived intentionality as a normative notion, where the normativity in question is of a linguistic kind. As he repeatedly says, the (internal) agreement between thought and reality that makes a particular subsisting state of affairs be the fulfillment of a certain intentional state is to be found in language, and language is intrinsically normative. Or, to put it more precisely, it is a rule of grammar that the intentional state that p is the state that p satisfies.

Yet the arguments that one can reconstruct on a textual basis in order to support this conception do not seem entirely convincing. On the one hand, one can exploit a straightforward argument for a linguistically normative intentionality to the effect that there cannot be a priority of thought over public language. According to this argument, if there were a prelinguistic thought, this could only be expressed by an impossible private language. Yet this argument basically relies on the implicit but controversial premise that any language is a normatively imbued structure. On the other hand, one can retrace an indirect argument for a linguistically normative intentionality to the effect that public language cannot derive its meaningfulness from a prelinguistic thought not only originally, but also intrinsically, endowed with intentionality, for there is no such thing: any attempt at singling out such a thought is doomed to fail, for it faces an infinite regress problem. Yet even this argument relies on a tacit premise claiming that thought has a physical design component, it is a (psychic) fact. This premise cannot be taken for granted if one accepts that intentional states have both original and intrinsic intentionality. These two elements – having both original and intrinsic intentionality – can indeed be simultaneously accommodated if one maintains that intentional states are abstract particulars having no physical basis.

At any rate, evicting intentionality from language as conceived by Wittgenstein, hence from normativity, does not show that Wittgenstein was wrong in thinking that meaning is an intrinsically normative notion. On the contrary, one can precisely combine a nonnormative view of intentionality with a normative conception of meaning: the meaning of an expression is what the intending of an intentional content becomes once that intending is expressed by correctly using the expression in question. Moreover, appearances notwithstanding, there is room for attributing to the later Wittgenstein not the ambitious thesis that public language precedes thought in general (so that intentionality is a linguistically normative property), but just the more cautious thesis that complex thoughts require language in order for them to be articulated.

1

It is hardly questionable that the main lesson one can derive from the later Wittgenstein concerning intentionality – the property of intentional states, or thoughts, of being about something (*intentionality of reference*) or of having a content that makes such states semantically evaluable, to be either true or false, either fulfilled or unfulfilled (*intentionality of content*)[1] – is that intentionality is a normative notion.[2] For according to him, intentionality eminently is a linguistic property and language is intrinsically normative. More precisely, the intentionality of content is basically a rule concerning the grammar of terms for intentional states' having a content. In Wittgensteinian's own words, 'like everything metaphysical the harmony between thought and reality is to be found in the grammar of language' (PG §112; see also PI §§429, 445), insofar as it is a rule of grammar that the intentional state that p is the state that p satisfies: 'the statement that the wish for it to be the case that p is satisfied by the event p, merely enunciates a rule for signs: (the wish for it to be the case that p) = (the wish that is satisfied by the event p)' (PG §112; see also PI §458). An analogous rule should be found for the intentionality of reference (cf. PI §462).

Yet, as we will see immediately afterwards, the two arguments that one may trace back to Wittgenstein in support of this lesson are not so convincing.

To begin with, there is a globally Wittgensteinian argument to the effect that language (in the sense of a publicly shared language) has priority over thought. As a result, this argument makes intentionality, *qua* property of intentional states, a linguistic property. The argument is an *ex absurdo* reasoning that, if there were a thought prior to language, this could be expressed only by a private language, a language understandable by its speaker only. But since there can be no private language for, as Wittgenstein has shown (PI §§258–61), one such language would break normativity by making following a rule collapsing onto believing that one is following such a rule, there is not even a thought prior to language.

Now, this argument is undoubtedly valid. Yet first of all, insofar as it incorporates one of the later Wittgenstein's most famous arguments, the private language argument (PLA), it relies on an implicit premise of that argument to the effect that language is intrinsically normative. For to say that a language whose normative character is threatened can be no language at all implies that normativity is an essential property of language. Yet people believing in different conceptions of language, such as Fodorian sustainers of the language of thought, the innate linguistic structure each human allegedly possesses, or Davidsonian fans of idiolects, languages which depend on their bearers for their identity, will not be impressed by that premise. For they will say that a public language is parasitic respectively on Mentalese or on idiolects for its own significance, and neither Mentalese nor idiolects are to be characterized as intrinsically normative.[3]

Clearly enough, if this were the only drawback of the argument, a Wittgenstein might reply that it is not so devastating: since that premise is precisely also a premise of the more specific PLA, and the PLA is one of the most fruitful cornerstones of Wittgenstein's later philosophy, why bother for the fact that that premise may sound controversial. The problem however is that, once one admits that that premise is controversial, it is no longer clear why one must stick to a previous premise in the argument, namely the thesis that if there were a thought prior to language, that thought could be expressed only by a private language. For once other, nonnormative conceptions of language are available, it is not to be taken for granted that the language expressing a previously existing thought is a private language. Incidentally, that premise did not always sound to Wittgenstein himself to be such a safe premise. This explains not only why he was not that sure to endorse it, as the preliminary texts to the *Investigations* show,[4] but also why when he endorses it in the *Investigations* he does so in a very oblique way (cf. PI §204).[5]

But there is another argument that a Wittgensteinian may exploit in support of the same result. This argument is independent of any consideration concerning private language and normativity. Yet as a side effect it may strengthen the previous argument, which, as we have just seen, relies on such notions. For, by indirectly weakening the idea that public language is parasitic for its significance on other meaning-relevant structures (linguistic or not), it reinforces that argument's assumption that language is intrinsically normative. According to this new argument, thought cannot have priority over public language for the former cannot lend significance to the latter, as the priority hypothesis presupposes. Hence, public language must find its significance in another way. Taken together, these two claims forcefully suggest that intentionality is again not basically a mental, but a linguistically normative property. Let me focus on these points in greater detail.

In the later phase of its reflection, Wittgenstein thoroughly criticizes the mentalistic thesis he himself apparently defended at the time of the *Tractatus*,[6] namely the thesis according to which public language gets its significance from underlying intentional states.

According to this thesis, first of all, thoughts have *original* intentionality, that is, they possess their significance by themselves, while language (like other public representations, such as pictures) has only *derived* intentionality, that is, it has significance only once it is interpreted, *meant, thought* as having a certain meaning. Moreover, thoughts also have *intrinsic* intentionality, that is, they cannot fail to have the significance they have; if they had no such significance, they would no longer be the same entities. On the contrary, language has intentionality only *extrinsically*, as is clearly shown by the fact that one and the same linguistic expression might have had a different meaning.

Although the two labels – original intentionality and intrinsic intentionality – are often used in the literature interchangeably as meaning the same thing, they express quite different notions, as I have just shown in the above paragraph. This is why, terminology notwithstanding, the first part of the above thesis can be defended independently of the second part, as the contemporary revival of the mentalistic picture clearly shows. That is to say, one may hold that intentional states have intentionality originally while failing to have it also intrinsically. According to this perspective, a thought would remain the same entity even if it failed to have the meaning it actually, and originally, has. This option is clearly favored by people championing a naturalistic account of intentionality, namely an account according to which intentionality is a property that can be studied by the natural sciences, in the same way as electricity or radioactivity,[7] although it does not imply endorsing that account.[8]

Yet in the interpretation favored by the early Wittgenstein, the two parts of the thesis cannot be split. To begin with, in that interpretation, for a public representation to have intentionality derivatively *means* to be paired with a hybrid entity, an entity-*cum*-meaning; namely, a two-faceted entity constituted both by the public representation itself and by its semantic interpretation. In Wittgenstein's own terminology, the former entity is called a *sign*, while the latter entity (the sign plus its interpretation) is called a *symbol* (TLP 3.32).

The main reason as to why having derived intentionality must be so understood is that this interpretation easily explains the phenomenon of semantic ambiguity, the fact that one and the same linguistic expression can have different meanings. For to have different meanings for an expression precisely means to be paired with different entities-*cum*-meaning, with different symbols, that share their first component – the sign – but differ in their second components, the interpretations of that sign (TLP 3.321, 3.323).

Once derived intentionality is understood in this way, it is easy to grasp why an entity having original intentionality must also have intrinsic intentionality. Since to lend meaning to an entity that originally has no meaning means to pair that entity with a hybrid entity, a certain entity-*cum*-meaning, and moreover that entity has that meaning as one of its components, hence as an essential part of its, then the entity-*cum*-meaning itself cannot but have intentionality intrinsically: if it hadn't the meaning it has, it would not be the same entity. But if this is the case, then the meaning-lending entity, the entity having intentionally originally, a thought, must have intentionality intrinsically as well as the hybrid entity; if it hadn't its (original) meaning that way, it could not lend it to anything else.[9]

As I anticipated above, taken this way the mentalistic thesis is deeply criticized by the later Wittgenstein. (In actual fact, Wittgenstein explicitly mainly criticizes the *experientialist* version of the thesis, according to which the meaning-lending intentional state is a *conscious* state, such as a certain mental state of imagination. Clearly, Wittgenstein focused his criticism against this version of mentalism for this is the version that has been widespread in philosophy across centuries, possibly the one that he himself defended in the earlier phase of his philosophy.[10] Yet it is also clear that Wittgenstein was prepared to extend his criticism to include other versions of mentalism that have gained popularity afterwards, such as *functionalist* mentalism, according to which an intentional state is a functional, or even a computational state, or *materialist* mentalism, according to which an intentional state is identical with a neurological state). First, intentional states are no *necessary* conditions for a linguistic expression to have meaning. Not only different users of a linguistic expression, or one and the same user at different times, may have different intentional states while using that expression, but it may also well be the case that one uses an expression meaningfully without having any intentional state associated with that use (cf. e.g., PI §35).

By itself, this failure does not rule out that linguistic expressions have derived intentionality, but at least it rules out that they *always* get their intentionality from intentional states. Yet second, the later Wittgenstein also holds that linguistic expressions *never* get their intentionality from intentional states. For intentional states are not even *sufficient* conditions for a linguistic expression to have meaning. Two people, or one and the same person at different times, may be in the same intentional state and still use a certain expression with different meanings (cf. e.g., PI §§139–40; II, iii, p. 177). This further failure already rules out the seemingly naturalistic idea that intentional states have original intentionality independently of their having intrinsic intentionality. Yet according to Wittgenstein it is intended to rule out the stronger claim that intentional states have both original and intrinsic intentionality. For, he says, the reason why intentional states cannot lend intentionality to public representations (as they are not sufficient conditions for the latter to have meaning) is that they are facts, brute facts – psychic facts, but this only means, facts that occur in someone's mind (or brain) – as much as those representations are. Now, Wittgenstein continues, suppose that an intentional state were taken together with a way to interpret the linguistic expression it is to be applied to – a method of projection, as he says (PI §141). This would not alter the situation by an iota: it may well be the case that *different* uses of one and the same linguistic expression are associated even to one and the same intentional state plus a method of projection (*ibid.*). Alternatively put, one may well form out of a psychic fact an entity-*cum*-meaning made by that fact plus a certain interpretation – the interpretation that psychic fact was initially supposed to be able to lend to a certain linguistic expression. Yet this not only shows that that psychic fact has no original intentionality, it is a meaningless brute mental or brainy fact, but it also presupposes that that fact would have to receive its intentionality from another, even more underlying, mental entity. Unlike the former, this latter mental entity should allegedly possess that intentionality intrinsically, in order to be able to lend its intentionality to that meaningless psychic fact so as to transform it in an entity-*cum*-meaning having intentionality intrinsically as well. But there is no reason as to why this even more underlying mental entity should not be in the very same predicament as the original intentional state was found to be – that is, that of being again a mere psychic fact (*ibid.*). So, an endless regress is started that shows that the mental way of ascribing intentionality to linguistic expressions is basically flawed: 'in the course of our argument we give one interpretation after another; as if each one contented us at least for a moment, until we thought of yet another standing behind it' (PI §201); '[B]ut [...] any interpretation still hangs in the air along with what interprets, and cannot give it any support. Interpretations by themselves do not determine meaning' (PI §198). So, it is not to be contested that it is the way a mental entity is applied that gives this entity its meaning (*a fortiori*, the meaning of the linguistic expression that mental entity is associated with); simply, that application is neither an original nor an intrinsic property of that entity (PI §141).

Thus, the last 'nonsufficiency' result not only again confirms that linguistic expressions have derived intentionality, simply they do not derive *at all* their meaning from underlying intentional states, but it also states that derived intentionality is an all-encompassing predicament that affects whatever has meaning: outer representations, but also inner representations such as intentional states, get their meaning by their being applied in a certain way. In this respect, Wittgenstein's position is close to Dennett's rejection of the derived/original intentionality distinction: whatever seems to possess intentionality originally turns out to receive its meaning from some other entity.[11] Yet unlike Dennett's instrumentalist position, Wittgenstein does not want to further stress that nothing genuinely possesses intentionality, for everything has intentionality only relative to a certain viewpoint (for instance, intentional states have intentionality only considered from an evolutionary perspective).[12] Rather, once they have meaning, both outer and inner representations genuinely have it, though derivatively; there are nevertheless further things that have only as-if intentionality, namely they have meaning only once they are regarded in the same way as those representations. For instance, animal thoughts or utterances (cf. PI §§360, 493–4). All in all, since as I said before for Wittgenstein nothing has original intentionality without also having intrinsic intentionality, the above result is intended to also show that nothing has both original and intrinsic intentionality.

By having the last consequence, that result for Wittgenstein shows that there is no primacy of thought over language. Therefore, it is indirect evidence that public language gets its meaning not by being thought but in another way. In actual fact, linguistic expressions (as well as inner representations, for that matter) get their meaning from elsewhere, not however from being vivified by some underlying entity, but by their being used in a certain way. 'Every sign *by itself* seems dead. *What* gives it life? – In use it is alive. Is life breathed into it there? – Or is the *use* its life?' (PI §432; cf. also BB, p. 4).

Yet that nothing has both original and intrinsic intentionality follows from the above reasoning only if one accepts an implicit presupposition that the meaning-lending entity allegedly having both original and intrinsic intentionality has a *factual* basis, is a worldly event. For the reason why a mental entity cannot be such a meaning-lending entity is precisely that any mental candidate for that role cannot actually play such a role for it has a factual basis: whatever has a factual basis cannot indeed have intentionality both originally and intrinsically. A psychic fact, which is a physical fact just placed in the mind (or brain), is just a brute fact like any other worldly fact, something to be studied by natural sciences. This is why the idea of a meaning-lending process for Wittgenstein is a no starter. It is indeed pointless to move from an outer representation (whether verbal or iconic), which is a worldly fact, back to an inner representation, which is another worldly fact as well. It is precisely because the outer representation is a mere fact that natural sciences investigate that it must be vivified by something in order for it to get meaning; but nothing mental can so vivify that representation, for it is another worldly fact as well. To be sure, its being a worldly fact would not prevent it from having mere original intentionality, as supporters of naturalized intentionality maintain – if this option were not already ruled out by the above 'nonsufficiency' result. Definitely, however, its being a worldly fact prevents it from being a superfact having both original and intrinsic intentionality. If it is a worldly fact, certainly, it can be transformed into an entity-*cum*-meaning having intentionality intrinsically, pretty much as it happens with a linguistic expression. But this precisely means that it must receive its intentionality from something else. In general, since the existence of a hybrid entity having intentionality intrinsically turns out of a transformation of an entity having intentionality derivatively (whether this latter entity is psychic or not), in order for that hybrid entity to exist there must be another effectively meaning-lending entity having intentionality both intrinsically and originally.

As we have seen, for the later Wittgenstein this predicament is where an endless regress starts. Yet in order for this regress to start, a possibility must be ruled out, namely that the ultimately meaning-lending entity has *no* factual basis, it does not rely on a worldly fact to be studied by natural sciences. Something nonfactual can have both original and intrinsic intentionality so as to lend meaning to entities originally failing to have it. By itself, however, Wittgenstein's argument above does not block this possibility.

Curiously enough, this theoretical possibility was definitely accessible to the early Wittgenstein. To be sure, when asked by Russell to give an example of a meaning-lending thought, he replied in a well-known letter that a thought is a psychic entity, whose elements are linked with the referents of the elements of the linguistic sentence to which that thought lends its meaning in the very same way in which these elements are so linked; a way that can be studied by a natural science such as psychology (NB, p. 130). In actual fact, this passage from the 1918 letter to Russell is standardly quoted by the defenders of the mentalistic interpretation of the *Tractatus*. Yet even supposing that Wittgenstein at that time was tempted by that interpretation, a moment's reflection would have shown him that a thought so conceived can only at most have original, but definitely not both original and intrinsic, intentionality. Indeed, a thought conceived as above well squares with present-day naturalist conceptions of intentional states.[13] That thought is actually a fact, an ultimately physical or anyway natural fact, precisely like the verbal sign to which it was supposed to apply. To be sure, as I said before, as much as outer representations that psychic entity may be transformed into an entity-*cum*-meaning. But this may happen only if there is another entity which is able to enact that transformation by having intentionality both originally and intrinsically, that is, only if there is an entity really playing the role a thought must play in Wittgenstein's framework.[14] Yet Wittgenstein's conceptual resources precisely yielded him another way of conceiving one such entity as a thought playing the above meaning-lending role in the way he himself favored, that is, as having intentionality both originally and intrinsically. That thought may well be the thought of the metaphysical subject, what Wittgenstein describes in the *Tractatus* as not belonging to the world insofar as it is the limit of the world (TLP 5.632), the condition for language to have significance hence for the world (better, for any possible world) to exist *qua* based on the fixed domain of the referents of the Tractarian names, the objects (TLP 5.62). This is to say, the metaphysical subject may well be seen as having thoughts endowed with both original and intrinsic intentionality that enable linguistic sentences (or even psychic facts, for that matter) to have meaning by thinking them, that is, by transforming them into entities-*cum*-meaning. Since the metaphysical subject does not belong to the world, his thoughts are not psychic, hence worldly, facts, but ultramundane (or better, borderline) entities.[15]

It is quite clear why the later Wittgenstein did not (re)consider this theoretical possibility. By making the meaning-lending entities thoughts of a metaphysical subject, this possibility explained something admittedly obscure – the idea that there is a meaning-lending entity having both original and intrinsic intentionality – by means of something even more obscure – the thinking activity of a metaphysical subject. This again collides with Wittgenstein's general ontological framework in the *Tractatus*. For the *Tractatus*, there can only be merely worldly facts, that is, the truth-makers of the only propositions that there are, the propositions of natural science (cf. TLP 6.41, 6.43, 6.53); thus, there cannot be superfacts endowed with both original and intrinsic intentionality not only *in* the world, but also *at its borders*, that is, in the realm of the metaphysical subject.

Yet that there are nonfactual thoughts playing that meaning-lending role can be accounted for without appealing to anything like a metaphysical subject. For instance, one may say that such thoughts are *abstract particulars*, that is, entities which non-spatiotemporally subsist so that they have no causal power and thereby not even a nature to be captured by physics, or anyway by natural sciences in general. Tropes – things like the sadness of Socrates and the redness of Plato – are precisely entities of this kind; moreover, as many have said,[16] things like Plato's thought that Socrates is sad or even more simply, Socrates' thought of Plato can well be ranked among such entities.[17] Granted, Wittgenstein's Tractarian ontology does not allow, at least explicitly,[18] for abstract particulars. Yet it remains that having shown that *no physically based entity* can have both original and intrinsic intentionality does not *ipso facto* mean that no entity whatsoever can have such features.

In the end, thus, one may well say that not only the first, but also the second argument one can trace back to Wittgenstein's later philosophy to the effect that there is no priority of thought over public language and that thereby intentionality is a linguistically normative property is not so convincing. For this argument also relies on an implicit premise that cannot be taken for granted, namely the premise stating the ultimately worldly nature of the meaning lending entity having both original and intrinsic intentionality.

As a result, suppose that one accepts Wittgenstein's claim that if anything possesses original and intrinsic intentionality it must possess both, for those features are inextricably intertwined.[19] Once one goes along that direction, there is room to conceive intentional states, and therefore intentionality, independently of public language, hence of normativity. As it indeed seems to be the case. Many different attempts at normatively accounting for intentionality notwithstanding,[20] at least as far as intentionality of reference is concerned, a thought is about what it is about, and there is nothing more to say, in particular nothing having to do with normativity: there is nothing like being the *right* (or *wrong*) target for one's thought.[21]

2

Let me now make two comments on this apparently negative evaluation of the later Wittgenstein's approach to intentionality. First of all, that Wittgenstein has not managed to prove that intentionality is a linguistically normative property does not say anything against his main tenet that the meaning of linguistic expressions is intrinsically normative. It may well be the case that, once one accepts that there is a gap between thought and public language, normativity is precisely what fills that gap. This may be easily seen in the simplest case, namely that concerning the relationship between intentionality of reference, a thought's being about something, and reference itself, a linguistic expression's designating something. If intentionality of reference is a property of nonlinguistic intentional states, as I said above there is nothing like being a *correct* target of one's thought: the thought is about what is about, and that's all. Yet the expression that one uses in order to express that thought may well be used to refer to the *wrong* referent, wherever what one has in mind, what one's thought is about, is just the *speaker's*, but not the *semantic*, referent of that expression (as when, by having Plato in mind, one utters 'Socrates').[22] For a linguistic expression's having a referent is already imbued with normativity: in a Wittgensteinian framework, for an expression to have a referent means to be used in accord with a certain rule that states that an application of that expression to a certain entity is the correct one (whereas the applications to other entities are the incorrect ones).[23] As a result, having for a thought a certain object (or a content in general) becomes the meaning of the corresponding expression of that thought once to apply the expression to that object (or with that content) amounts to the *correct* use of that expression.

Moreover, the received view notwithstanding, it must not be taken for granted that the later Wittgenstein defended the thesis that is commonly ascribed to him and that I have hitherto attributed to him as well, namely the claim that public language has priority over thought *in general*. This suspicion has not to do with a generically methodological note of caution about the later Wittgenstein, according to which in the later phase of his reflection Wittgenstein refrained from endorsing philosophical theses at least when they are controversial.[24] Rather, it has to do with the fact that Wittgenstein himself was explicitly cautious on this point. If one wants to accurately ascribe him a position on this point, one should rather ascribe him the position – that has recently gained popularity among cognitive scientists[25] – of those who claim that only *complex* thoughts require (public) language in order for them to be articulated:

> One can imagine an animal angry, frightened, unhappy, happy, startled. But hopeful? And why not?
>
> A dog believes his master is at the door. But can he also believe his master will come the day after tomorrow? – And *what* can he not do here? – How do I do it? – How am I supposed to answer this?
>
> Can only those hope who talk? Only those who have mastered the use of a language. (PI II, i, p. 174; cf. also §650)[26]

By 'complex thoughts' one must here mean thoughts whose conceptual constituents need a linguistic expression in order for them to be individuated. For example, because they exercise their inferential role in a net of inclusions and exclusions – as is the case with the concept of *the day after tomorrow* Wittgenstein mobilizes in the above example (so that one may grasp that if April 1 is the day after tomorrow, it is neither tomorrow nor today)[27] – or because a verbal manifestation of what the concept denotes is required in order to tell that concept from other similar ones – as it may well be the case with sensation concepts ('you have learned the *concept* "pain" when you learned language': PI §384).

Thus, the later Wittgenstein's detailed position on this matter seem to be that *not* all concepts, hence not all thoughts, need language in order for them to be articulated. If this is the case, it is not so obvious that for him intentionality is just a linguistically normative property. Now, as we have seen before, the fact that intentionality has nothing to do with public language seems to be well proven by its failing to have a normative character. Thus, if the present interpretation of Wittgenstein's *global* position on intentionality were correct, it would indeed give us a theoretically welcome result.[28]

Notes

1. For these labels, cf. Kim (1996, p. 21).
2. If this really exhausts Wittgenstein's position on the matter, however, it will be thoroughly scrutinized in the next Section.

3. As to the difference between Mentalese and a private language on this issue, cf. Marconi (1995, pp. 112–14). For a deeper scrutiny of this point, cf. Calabi-Voltolini (2005).
4. For a thorough discussion of this point, cf. Pears (1988, pp. 372–9).
5. A more explicit endorsement of this thesis is in RFM, VI §32.
6. At least according to some of the most influential interpretations of the *Tractatus*; cf. for example, Malcolm (1986, p. 66). For the time being, I remain neutral as to whether the thesis must be meant in a mentalistic way, according to which the meaning-lending intentional states are psychic events in the mind of an empirical subject, or in a nonmentalistic way, according to which those states are not such events. I will deal with this problem later in the text.
7. Cf. for example, Dretske (1988, 1990, 1995), Fodor (1987, 1990).
8. For a not necessarily naturalist endorsement of the 'originality without intrinsicness'- thesis, cf. for example, Kriegel (2007).
9. This 'no-splitting'- position is nowadays defended by Searle, apparently along the same lines: 'the mind imposes Intentionality on the production of sounds, marks, etc., by imposing the conditions of satisfaction of the mental states on the production of the physical phenomena' (Searle, 1983, p. 164); 'just as the conditions of satisfaction are internal to the speech act, so the conditions of satisfaction of the Intentional state are internal to the Intentional state' (Searle, 1983, p. 11).
10. We will see later to what extent this exegetical conjecture is correct.
11. Cf. Dennett (1990).
12. Cf. Dennett (1987, 1990). For a clear distinction between the notions of derived intentionality and 'as-if' intentionality, cf. Brandl (1996).
13. Cf. McDonough (1994).
14. Incidentally, this point eludes many interpreters of the early Wittgenstein, who defend a mentalistic interpretation of the notion of thought. For them, it may well be the case both that a thought is just a psychic fact and that that thought can be also seen as intrinsically possessing an interpretation. Cf. for example, McDonough (1994). To repeat, the point is that, if there is a psychic entity and that entity can also be seen as intrinsically possessing an interpretation, then there must be *another* entity that has intentionality both originally and intrinsically in order for that psychic entity to be so seen, that is, in order for that entity to be transformed into a hybrid entity having intentionality intrinsically. The thought must be *this other* entity. To be sure, the early Wittgenstein made things obscure by labelling 'thought' also the hybrid entity: 'a thought is a proposition with a sense' (TLP 4). This sentence should however always be read along with the immediately previous sentence: 'a propositional sign, applied and *thought out*, is the thought' (TLP 3.5; my italics), that clearly shows that there must be thought over and above the thought that coincides with a hybrid entity.
15. Hacker (1972, pp. 47–8, 77) limits himself to allowing for the metaphysical subject to play the meaning-lending role that projects any fact whatsoever (a verbal sign, but even a psychic fact) to a state of affairs possibly subsisting. Yet in playing that role the metaphysical subject must *think* the to-be-projected fact. For, as Wittgenstein says, 'the method of projection is to *think* of the sense of the proposition' (TLP 3.11; my italics; see also the

already quoted 3.5). So, Hacker's interpretation of the early Wittgenstein's theory of thoughts must be actually developed in the way proposed here. Also Kenny's (1981, p. 146) interpretation of this point, which in this respect ascribes to the metaphysical subject meaning-acts of the will, goes in this direction.

16. For thoughts as (relational) tropes, cf. for example, Mulligan and Smith (1986), Bacon (2002).
17. For a development of such a position, cf. Sacchi and Voltolini (2009).
18. One might indeed hold that a thought considered along the lines of TLP 3, namely as the abstract core of any picture whatsoever, that is, as the logical role that allows a picture to possibly represent all the states of affairs sharing its logical form – cf. on this Frascolla (2007) – is precisely tantamount to an abstract particular.
19. To be sure, this thesis has been recently disputed in the contemporary philosophy of mind. Yet the reasons against it are far from conclusive. What one basically finds in the literature is a naturalist prejudice towards that thesis. Yet independently of the issue of its correctness this prejudice is grounded only if a defender of that thesis also maintains that the intentionality an intentional state possesses both originally and intrinsically is a nonnatural property of that state. Cf. on this my Voltolini (2005).
20. Cf. for example, Putnam (1988, pp. 11–15), Searle (1992, p. 51), Brandom (1994, 2000).
21. At most, intentionality has normative implications, but it is not intrinsically normative. For more on this, cf. Calabi and Voltolini (2005).
22. Cf. again Calabi and Voltolini (2005).
23. Incidentally: nothing shows that the later Wittgenstein was against partially endorsing a normatively adapted direct referentialism on meaning. For not only direct referentialism is quite compatible with a normative account of meaning – it may well be the case that directly referential expressions are those which are used in accord with rules of the form '"R" refers to R' – but nothing of what Wittgenstein says in the *Investigations* actually goes against at least a partial adoption of direct referentialism. Wittgenstein's famous 'Moses' – example (PI §79) is intended not to support the so-called 'cluster version' of descriptivism, as is commonly repeated after Kripke (1980), but the subtler claim that, *if* one sticks to descriptivism, one may find in the case of 'Moses' another example in which one preserves the truth of a sentence by changing its meaning. See also Hacker (1996b, p. 223). More positively, the famous passage of PI §43 literally says that sometimes the (normatively interpreted) meaning of an expression is defined by pointing out the expression's bearer (as it happens in a referential rule of the above kind), for this is just a particular way for the 'meaning is use'– slogan to hold – a slogan, recall, that holds only when 'meaning' means *linguistic meaning* and not as regards *other* uses of that particular expression (as e.g., when we say 'the meaning of life'), as the first sentence of that well-known paragraph says.
24. 'If one tried to advance *theses* in philosophy, it would never be possible to debate them, because everyone would agree to them' (PI §128).
25. For an overview, cf. for example, DeCaro, Ferretti, and Marraffa (2007).

26. Once one so reads the passage quoted in the text, *Investigations'* famous §25, that has always been interpreted as supporting the 'language-over-thought'-priority thesis, gains a new light: 'It is sometimes said that animals do not talk because they lack the mental capacity. And this means: "they do not think, and this is why they do not talk." But – they simply do not talk. Or to put it better: they do not use language – if we except the most primitive forms of language. – Commanding, questioning, storytelling, chatting, are as much as a part of our natural history as walking, eating, drinking, playing.'
27. That this is the way Wittgenstein remark has to be interpreted, rather than merely in the generic sense that one needs language in order to have a sense of the future, is clear. For otherwise that remark would be empirically false, as has been clearly shown by Bandura's psychological studies on dogs' learning how to self-regulate in not eating whatever is contained in a fridge they manage to open. Cf. Hutchins, Hockenbury and Hockenbury (2002).
28. I'm very grateful to Diego Marconi for his comments to a previous version of this chapter.

Part II

Meaning

5

Does Language Require Conventions?

Hans-Johann Glock

1 Introduction

According to David Lewis, it is a 'platitude – something only a philosopher would dream of denying' – that language is governed by conventions (1983, p. 166). Noam Chomsky, Donald Davidson, and their followers beg to differ. In recent years, they have fervently attacked the idea that rules, conventions, or a common language have a useful role to play in an account of language.

Although the question has been discussed mostly by reference to Chomsky and Davidson, it has wider implications. For it epitomizes several fundamental controversies in contemporary philosophy of language and beyond.

The first is the conflict between naturalists and normativists. Naturalists believe that there is no genuine knowledge outside natural science (epistemological naturalism), and that the phenomena studied by natural science (roughly, matter, energy, spatiotemporal objects and events) exhaust what there is. Against this position, philosophers since Kant have objected that logic, knowledge, and language depend on norms that cannot be captured by the factual descriptions and causal explanations of the natural sciences.

The second is the conflict between individualistic and communitarian perspectives on language. According to individualists, the fundamental linguistic phenomenon is the idiolect, the language of an individual. According to communitarians, the fundamental phenomenon is that of a shared public language. They treat language as a social practice or a kind of institution which is more than the sum of individual speakers or idiolects.

Finally, this debate has repercussions beyond philosophy, notably for our understanding of culture and society. Individualism is hospitable to the idea of an extremely multifaceted and fragmented society. We not only hold diverse beliefs and values, we need not even share a common language. Communitarians, by contrast, insist that we must share at least a linguistic medium in which to express our conflicting viewpoints. And because they treat language as an intersubjective practice, they can acknowledge and deplore phenomena such as 'language death,' the increasingly common disappearance of natural languages such as Cornish. According to individualists, on the other hand, so called language death is merely a quantitative shift in patterns of verbal behaviour, no more significant than, for example, the death of an individual speaker (see Glock, 2004).

In this chapter I shall disregard these wider implications of my title question. I shall merely be concerned with defending an affirmative communitarian answer against the new individualist orthodoxy. I shall further confine myself in two respects. First, I shall concentrate on the question of whether language requires *lexical* or *semantic* conventions, norms which fix the meaning of particular expressions and largely disregard the question of whether there are syntactic or pragmatic conventions. Secondly, I shall confine myself to the following brief remarks about one currently influential version of individualism, namely the psychological one which denies that language has any essential connections to communication.

According to Chomsky, a language is a 'system of knowledge,' a 'state of the language faculty,' which itself is an internal 'component of the mind/brain' of individuals (1988, p. 60). He calls languages in this sense 'I-languages,' and contrasts them with languages in the everyday sense, which he calls 'E-languages.' This model is not just individualistic but psychological. An I-language is not the idiolect of an individual speaker; it is an internal state which causally underlies that idiolect.

Chomsky's psychologistic conception entails that language has no essential connection with communication, and *a fortiori* with shared norms of communication. It has the unpalatable (if openly acknowledged) consequence that English, Mandarin and Xhosa, for example, are not languages and strictly speaking do not exist. This deprives the study of language of its ultimate topic, which consists in patterns of intersubjective linguistic behaviour. Even those linguists who are interested in its ultimate, presumably neurophysiological, causes at the individual level start from those observable patterns of intersubjective behaviour as their *explanandum*. And without that explanandum, the postulation of Chomsky's I-languages is not just speculative, but downright pointless. To explain the behaviour of individual speakers we may have to regard it as the exercise of a specific language faculty, and that faculty may depend on specific areas in the brain. But the faculty itself is a set of complex abilities possessed by speakers. To identify it with either the brain or a state of the brain is to confuse an ability with its vehicle, the physical mechanism or structure in virtue of which an individual has an ability (see Kenny, 1989, ch. 5). And to deny the existence of the exercise of the ability by appeal to the vehicle is to saw off the branch on which anyone talking about faculties, capacities or abilities must be sitting.

Furthermore, Chomsky's denial of the link between language and communication has stretched the credulity even of some of his admirers. It creates insurmountable difficulties for explaining the origins of language. There is no conceivable evolutionary explanation of why our hominid ancestors should have developed biologically costly neural structures simply in order to soliloquize, especially since as social primates they had an immediate and pressing need for media of communication and coordination (see Greenspan and Shanker, 2004). Chomsky describes 'the language faculty' as an organ. But an organ, especially one that is not a physiological unit, had better had a function. Chomsky seems to acknowledge this point, since he describes language as a means for expressing thoughts. Yet it is difficult to see what principal purpose the expression of thoughts could serve, if not that of communicating.

Davidson represents a more promising type of individualism. He acknowledges the link between language and communication, while arguing that linguistic communication does not presuppose conventions and cannot be explained by reference to shared rules or conventions. Nevertheless I hope to show that even this variant of (semantic) individualism fails.

The communitarianism I defend is inspired by Wittgenstein and some of his followers, and shares some assumptions with Davidson. In Section 1, I show how two current debates, about the normativity of meaning and the conventionality of language respectively, both emerged from Wittgenstein's philosophy of language. In Section 2, I address the question of what a convention is, defining it as a shared, arbitrary rule. Section 3 defuses a venerable argument against the appeal to conventions in an account of language, namely that conventions presuppose rather than explain language. In Section 4, I argue against current claims that there are no lexical conventions because there are no lexical rules.

That meaning is arbitrary is granted by individualists. Ironically, however, they try to turn this feature against communitarianism. Section 5 defuses this threat, arguing in particular against the idea that each one of us speaks her own idiolect. In Section 6, I turn to the question of whether lexical rules must be shared, which I answer in the affirmative. The final two sections diagnose where individualism has gone wrong.

2 Wittgenstein and semantic normativity

The idea that language is a *rule-guided activity* is as old as the study of language. It underlies the discipline of *grammar*, whether in its traditional or modern form (descriptive and generative grammar). It was given an additional boost by the development of *modern logic* at the turn of the twentieth century: logical calculi evolve around formation rules (syntax) and transformation rules (rules of inference). Nevertheless, within analytic philosophy, the idea that there is an irreducibly normative dimension to linguistic meaning goes back to Wittgenstein. In a text from the early thirties which is based on dictations by Wittgenstein, Waismann wrote:

> Of course words impinge on human life, of course they produce various effects. What we want to say is merely this: we do not regard language from the perspective of a mechanism, but from that of a calculus, a procedure according to norms.... In the picture of a psychic mechanism, in which language pulls the lever that moves the body of the hearer, the concept of meaning does not yet appear. There is only room for this concept if one regards language from a normative perspective. Indeed, the concept of 'meaning' is felt to be an alien element in the considerations of natural science. And our analysis shows why this is so. (1976, pp. 184–5, 190–1)

In this passage we find three fundamental ideas of contemporary normativism: first, meaning is essentially normative; secondly, this normativity defies reduction to the kind of causal relations investigated by natural science; thirdly, the reason is not that meaning or meaningful language involve supernatural phenomena; rather, the normative dimension reveals itself to a special *perspective*, a unique yet indispensable way of looking at phenomena that can also be described scientifically.

Nonetheless it was only through Kripke's influential interpretation of Wittgenstein's so called 'rule-following considerations' that semantic normativity came to occupy the centre of philosophical attention. In the wake of Kripke the debate has shifted significantly relative to its Wittgensteinian roots (see Glock, 2005, pp. 220–8). At present it features a whole barrage of normativity claims. All of them involve the idea that a certain phenomenon which is semantic in a highly general sense – including contentful thought as well as meaningful speech – entails the existence of certain normative phenomena. Two claims which are important to the issue of conventions are what I have called *bare normativity of meaning* (BNM) and *rule-based normativity of meaning* (RNM). BNM maintains that if a word *w* is meaningful, then there must be conditions for its correct application. According to RNM, meaningful expressions entail not just the existence of correctness conditions, but of rules of use, because the correctness conditions of BNM are laid down by such rules.

RNM is rejected by some who subscribe to BNM, such as Davidson and Boghossian, and accepted by speech act theorists and many Wittgensteinians. At the same time other followers of Wittgenstein acknowledge that he insisted on a normative dimension to meaning (BNM), while denying that he regarded language as rule-governed (RNM). They also suggest that RNM is, in any case, incorrect (e.g., Cavell, 1962; Hanfling, 1989; Rundle, 1990, pp. 5–8, 33–4, 64–9). According to these 'unruly' Wittgensteinians, his comparison of language to rule-following activities such as games should be seen as a dispensable, if not misleading, heuristic device, which perhaps betokens a schoolmasterly attitude he imbibed while teaching in Lower Austria in the 1920s.

This view is correct insofar as Wittgenstein became suspicious of his own earlier picture of linguistic rules. However, his reaction was *not* to abandon the notion of a linguistic rule, but to *clarify* it. This is precisely the role of the later discussion of rule-following.

In the *Tractatus* linguistic rules form 'logical syntax,' a complex calculus of inexorable norms hidden behind the surface of natural languages (see 3.3ff., 4.002, 5.4541, 5.5562). By the mid-thirties Wittgenstein had moved away from this 'calculus model' of language (see Glock, 1996, pp. 67–72). He denied that rules can guide linguistic behaviour and determine what it makes sense to say without being known to us. He also rejected the idea that rules are inexorable or independent of human activities, while retaining the idea that human practice is in many respects rule-governed. In particular, Wittgenstein changed his conception of linguistic rules by comparing language no longer to a calculus of hidden and rigid rules, but to a game. *That* comparison, however, is not just a dispensable heuristic device. Wittgenstein continued to insist that linguistic understanding involves mastery of techniques concerning the *application of rules*. 'To follow a rule, to issue a report, to give an order, to play a game of chess, are *habits* (customs, institutions). To understand a sentence means to understand a language. To understand a language means to be master of a technique' (PI §199).

We need to avoid not just the calculus model of hidden rules, but also an intellectualist picture of rule-following, according to which it necessarily involves the consultation and interpretation of rule formulations. For this picture leads to Wittgenstein's notorious regress: to interpret the initial rule formulation, we replace it by another one, which in turn needs to be interpreted, and so on. As a result, we would end up with a 'paradox:...a rule could not determine a course of action, since every course of action could be made out to accord with the rule.' Does Wittgenstein conclude that following a rule is strictly speaking impossible, as rule-sceptical and unruly readers have suggested? Not at all! He immediately goes on to diagnose the paradox as based on a 'misunderstanding.' We

> posit one interpretation after another; as if each one contended us at least for a moment, until we thought of yet another standing behind it. What this shows is that there is a way of grasping a rule which is not an interpretation, but which is exhibited in what we call 'obeying a rule' and 'going against it' in actual cases. (PI §201)

On the basis of this down-to-earth conception of rules, Wittgenstein continued to insist on the link between grammatical rules and meaning. In 1930 he remarked in a lecture:

> What justifies us in using any particular word? Suppose I say 'This gown is black.' The word black is arbitrary in one sense; another sound or scratch would serve. And the correlation of the word 'gown' to a particular object is in itself arbitrary and has no consequences. But if a proposition is to have sense we must commit ourselves to the use of the words in it. It is not a matter of association; that would not make language work at all. What is essential is that in using the word I commit myself to a rule of use. A world only has meaning in a grammatical system, and what characterises it is the way in which it is used. (LWL p. 36)

To some ears, talk of a grammatical system may still smack of an overly rigid conception of linguistic rules. But in stressing *commitment to a practice*, this passage already avoids both the idea of unconscious rule-following and the intellectualist mistake. Furthermore, right to the end of his career Wittgenstein expressed his central insight that meaning is a matter of use through invoking the idea of rules. '... A meaning of a word is a kind of employment. That is why there exists a correspondence between the concepts "rule" and "meaning"' (OC §§61–2).

Small wonder, then, that Wittgenstein maintained that 'following according to the rule is FUNDAMENTAL to our language-game' (RFM p. 330). Rule-following plays a central role in Wittgenstein's overall oeuvre because of two abiding convictions: first, language is a rule-guided activity; secondly, the a priori status of logic, mathematics and philosophy derives from such rules. More specifically, philosophical problems are to be resolved by attending to rules of use:

> We are interested in language as a procedure according to explicit rules, because philosophical problems are misunderstandings which must be removed by clarification of the rules according to which we are inclined to use words. We consider language from one point of view only. (PG p. 68; see also BT pp. 424–5)

But why is this 'one point of view only'? First, because language can also be viewed from a nonnormative, purely causal 'perspective of a mechanism,' as the Waismann passage states (see also PI §§491–3). Secondly, our linguistic practice is rarely guided by explicitly stated or rigid rules. As long as this is kept in mind, however, nothing prevents us from explicating rules for philosophical purposes (AWL pp. 47–8; Z §467). And the same holds for stressing the rule-guided character of language in general. Accordingly, there are overwhelming reasons for ascribing to Wittgenstein not just BNM – the idea that meaning is normative – but also RNM – the idea that meaning is linked to rules. The question is whether he provides us with compelling arguments in their support.

3 Regularities, rules and conventions

I have discussed that question elsewhere (Glock, 2008). The current essay is devoted to another normativity claim, namely that meaningful language requires the existence of conventions. To assess this claim, we naturally need to clarify what is meant by a convention. I shall take for granted that a convention is a kind of norm or rule. Norms in the relevant sense differ from mere regularities or natural laws in two interrelated respects.

First, they do not merely describe or explain, but set standards, in particular standards against which behaviour can be assessed as correct or incorrect. If planets turn out to follow paths other than elliptical ones, it is Kepler's Laws rather than the movement of the planets which needs revision. By contrast, if I drive on the right hand side it is my driving rather than the Traffic Code which is at fault.

Secondly, norms provide reasons for action. Although rule-following presupposes a regularity in behaviour, there is a difference between rule-guided and merely regular behaviour. If an agent follows a rule in Φing, the rule must be part of his reason for Φing, and more specifically, for Φing in a particular manner.

What then sets conventions apart from other rules? According to Lewis (1969), conventions are communal regularities of behaviour that arise within 'coordination problems,' problems of arriving collectively at a particular course of action in situations where other courses of actions are equally feasible, provided that they are adopted collectively. The regularity is sustained not just by complex beliefs and preferences concerning those who are party to the convention, but also by a common knowledge that they have these beliefs and preferences.

This account does not capture what we standardly mean by a convention. For one thing, many conventions, for example some rules of etiquette and custom, do not solve problems of coordination (Gilbert, 1989, ch. 6). For another, Lewis demands far too much from those party to a convention. He is prepared to bite the bullet of accepting that 'children or idiots' cannot be governed by conventions (1969, p. 51). But the problem runs much wider. By and large, people do not entertain the complex beliefs and desires concerning members of large and often unspecified groups around which his definition revolves (see von Savigny, 1983, §8). And while Lewis goes wrong in requiring from the subjects complex beliefs and desires, Gilbert goes wrong in lumbering them with explicit joint commitments. Both are overly intellectualistic. They ignore the Humean insight into the importance of custom and habit. The substratum of conventions is regularities of behaviour that have become so entrenched and internalized as to give rise to normative expectations.

For these reasons I favour a weaker definition of conventions. It starts out from a proposal by Herbert Hart (1961, pp. 54–6). Following Hart, in a community C a behavioural regularity R is a *shared rule* iff

(a) it is rare for members of C to deviate from R,
(b) if members of C deviate from R, they are subject to sanctions, including the verbal sanction of being criticized,
(c) these sanctions are generally accepted by members of C.

Hart adds an additional requirement. Members of C regard R from an 'internal point of view,' that is, they avoid deviating from it not just for reasons of prudence, notably for the sake of avoiding sanctions. But as von Savigny has argued, this requirement is too strong (1988, ch. 2). Although as a matter of fact no rule-following practice may be stable if compliance with it is based exclusively on sanctions, this is not part of the concept of a rule that is in force within a certain community.

However, as I stressed above, behaviour is guided by rules only if these provide the agent's reasons for proceeding as they do. Therefore we need to preserve one element of Hart's additional requirement:

(d) members of C abide by R for a reason (which can either be the communal regularity itself or the sanctions associated with deviations).

Even if (a) – (d) provide an acceptable definition of the concept of a shared rule, however, they do not capture the concept of a convention. For they apply equally to shared *prudential* rules, such as the rule not to go out into the rain unprotected. For this reason I propose to define conventions as shared rules that satisfy an additional requirement:

(e) it would have been feasible for members of C to proceed according to a different regularity R′ without incurring substantial disadvantages.

As Lewis noted, the terms 'convention' and 'conventional' carry a connotation of arbitrariness: there is no such thing as 'the only possible convention' (1969, p. 70).

It is important to note, however, that the arbitrariness of conventions has to be treated with care. First, it does not preclude some conventions from being superior to others. Systems of measurement are arbitrary, and yet the metric system is clearly superior to the imperial one. Secondly, to say that a regularity R is arbitrary in the sense of (e) is not to say that it would be convenient or even feasible to alter it. It may be unfeasible for most of the world to switch from driving on the right to driving on the left, yet this is a convention nonetheless. That is why (e) speaks counterfactually about the past. Thirdly, mere feasibility or possibility (as Lewis has it) is not enough. It is feasible in principle to *abandon* many of the rules of the criminal code, yet these are clearly not conventional in the same sense in which some other legal rules are, not to mention the rules of etiquette. Fourthly, we have to distinguish the convenience or feasibility of other regularities from the convenience or feasibility of other sanctions. It would not be convenient or even feasible to impose the death penalty for every breach of etiquette, yet it might be convenient to adopt a different etiquette. Finally, not all arbitrary shared rules would commonly count as conventions. Consider for instance the absurd rules dictated by tyrants – for example, having to sing the International at the end of every meeting. It may seem that these are not conventions, because they are accepted solely because of fear of sanctions. And this may further suggest that at least with regard to the term 'convention,' we need to reinstate something like Hart's internal point of view. But while I agree that we would tend to speak of dictates rather than conventions here, I do not think that we would regard the latter term as inappropriate.

Given these provisos, my definition of a convention might be summarized by saying that a convention is a *shared, arbitrary rule*. It is a *rule* because it provides reasons for regular behaviour either directly or because of sanctions – (d). It is a *shared* rule because conformity is widespread in a community, because deviations are subject to normative reactions, and because these reactions are generally accepted – (a) to (c). And it is *arbitrary* because a different rule might have been adopted – (e).

4 Is the appeal to conventions circular?

Having dealt with the notion of a convention, let us now go back to our core topic of semantic or lexical norms. On the assumption that there are lexical rules and that these rules are arbitrary, the debate between individualist and communitarian turns to the question of whether these rules must be shared between speaker and hearer, or, more generally, between the members of a linguistic community. This is not the problem debated by readers of Wittgenstein's private language argument, namely whether there could be a language that is not or could not be communicated to others. For we are *ab initio* concerned with cases of communication. The question is rather whether an appeal to conventions can elucidate the nature of linguistic communication.

There is an immediate objection to this idea. Explaining language by appeal to conventions is circular, because language is needed in order to set up conventions. In Russell's words:

> We can hardly suppose a parliament of hitherto speechless elders meeting together and agreeing to call a cow a cow and a wolf a wolf. (1921, p. 190)

Davidson makes the same point in a more assuming manner:

> Beliefs, desires, and intentions are a condition of language, but language is also a condition for them. On the other hand, being able to attribute beliefs and desires to a creature is certainly a condition of sharing a convention with that creature; while, if I am right in what I have said in this chapter, convention is not a condition of language. I suggest, then, that philosophers who make convention a necessary element in language have the matter backwards. The truth is rather that language is a condition for having conventions. (1984, p. 280, see 27)

In this passage, Davidson argues for a Quinean conclusion, but in a distinctive fashion. It is *prima facie* plausible to hold that conventions presuppose language, since they must be expressed linguistically. Unlike Davidson (2001, pp. 113–14), however, Quine acknowledged that there might be *implicit* rules that guide our behaviour without having been enunciated or being consciously consulted. His complaint is that the appeal to such implicit rules lacks explanatory force (see Glock, 2003, 86–95 for an attempt to resist this complaint). Davidson's argument, by contrast, runs as follows:

(P_1) All conventions (implicit ones included) presuppose beliefs and intentions about the beliefs and intentions of others

(P_2) Beliefs and intentions about the beliefs and intentions of others presuppose language

(C) All conventions presuppose language.

The merits of (P_1) depend on the concept of convention. It does hold for the definition by Lewis that Davidson is building on. But it is implausible for the weaker conception I have favoured, which demands only that the participants be capable of intentional action and of beliefs and attitudes towards the *behaviour* of others. For sanctions can in this context be understood without reference to the intentions of those who impose them, namely as adverse reactions to a deviation from a regularity. In fact, language probably arose when the regular use of signals started to provide individuals with reasons for verbal behaviour and then became subject to normative reactions and sanctions.

In any event, I hold that (P_2) is wrong. *Pace* Davidson, nonlinguistic creatures can have not just beliefs and intentions, but also beliefs and intentions concerning the beliefs and intentions of others. A majority of animal psychologists maintains that great apes have such beliefs (what they call 'a theory of mind'), since they pass some false belief tests and engage in tactical deception like the hide and peek strategy. And many of the most promising genetic theories of both language evolution (Tomasello, 1999; Rizzolatti and Arbib, 1998) and language acquisition (Bloom, 2002) take possession of a theory of mind to be a prerequisite for language. Admittedly, other empirical researchers have questioned these attributions of a theory of mind to great apes. In any event, however, Russell's and Davidson's philosophical circularity objection founders. (P_2) is not a conceptual truth, since we can clearly and easily conceive of nonlinguistic hominids that display wide-ranging beliefs about the mental lives of their conspecifics (see Glock, 2000).

5 Are there lexical rules?

Given this understanding of rules and conventions, language must be governed by rules in order to be governed by conventions. Prima facie, this is not much of a liability for communitarianism. The view that language is governed by rules is at least as much of a commonplace as that language is governed by conventions. It is enshrined in disciplines like grammar (irrespective of whether it be prescriptive, descriptive or generative), logic (irrespective of whether Aristotelian or Fregean) and formal semantics. It is equally shared by philosophers who regard language not so much as a formal system, but as a form of behaviour – notably speech act theorists and Wittgensteinians. Thus John Searle writes:

> Speaking a language is to engage in a rule-guided form of behavior. To put it more briskly, talking is performing acts according to rules. (1969, p. 22; see also Baker and Hacker 1985, p. 154:
>
> For all that, individualists like Chomsky and Davidson have attacked the very idea that language is rule-governed, including the idea of lexical rules. Thus Davidson has recently written that the yearning for [linguistic] norms is a nostalgic hangover from the dependence on a Platonic conception of meaning (1994a, p. 145; see also Bilgrami, 1993).

But flattery will get them nowhere. It can be shown that the very same individualists rely on the picture of language as a system of formal rules (see Glock, 2003, pp. 250–2). More importantly, BNM marks a straightforward normative dimension to meaning. Because of what words mean, some of their applications are correct while others are incorrect. For example, it is correct to call an unmarried, unwidowed, adult male a bachelor, a male duck a drake, or this folder black; it is incorrect to call a married man a bachelor, a female duck a drake, or this diary white, because of what the terms 'bachelor,' 'drake' and 'white,' respectively, mean. To put it differently, meaningful expressions possess conditions of correct application or of correct use (see Boghossian, 2002b, p. 148).

There are two individualist objections against this picture. The first is that we are dealing here merely with 'descriptive facts' concerning linguistic behaviour (see Davidson, 1994a, p. 146; Glüer, 2002). Now, it is true that a proposition like:

(1) In English, it is correct to apply 'drake' to an object x iff x is a male duck

states a fact about the behaviour of Anglophones. But it does so only because the Anglophone speech community accepts a certain norm, a standard of correctness. We must distinguish between (1), which is an empirical proposition to the effect that a community follows a norm, what von Wright calls a 'norm-proposition' (1963, viii), from the expression of the norm itself, as in:

(2) It is correct to apply 'drake' to an object x iff x is a male duck or a standard definition like

(3) A drake is a male duck

Only (2) has a metalinguistic form, that is, it mentions a linguistic expression. Nevertheless, both (2) and (3) are standardly used to explain what the term 'drake' means. And because they function as explanations, they also have a normative role vis-à-vis the use of the term 'drake.' They lay down under what conditions the term 'drake' can be *meaningfully applied* to an object x. Whether the term *actually applies* to x then depends on whether x satisfies these conditions (for other points in defence of semantic normativity see Glock, 2005; 2009; Calabi and Voltolini, 2005; Whiting, 2007).

It is difficult to deny that 'correct' and 'incorrect' are normative terms, terms of appraisal. However, the individualist has a second objection in store, one directed against RNM. Even if (2) has some kind of normative force, it is not a rule, since it does not guide behaviour (see Glüer and Pagin, 1999; Stainton, 1996, pp. 139–42). It does not, for example, oblige or even entreat us to apply the term 'drake' to an object in a particular situation.

This second complaint is based on an unduly narrow conception of what counts as a rule. It ignores the 'definitory aspect of rules' (Baker and Hacker, 1984, p. 259). Some rules do not prescribe a certain form of behaviour, but instead lay down what a thing must be like to satisfy a certain description. Prime examples are the beloved EU norms concerning, for example, what counts as a sausage. This point is connected to a well known distinction between regulative and constitutive rules (e.g., Searle, 1969, ch. 2.5). The former specify how one should optimally pursue an activity that can be specified independently of the rule; the latter are partly definitive of the activity in question. What goes on at dinner parties can be described as eating independently of the rules of etiquette. By contrast, what goes on when people move certain pieces across a board cannot be described as castling, checking, mating, or, more generally, as playing chess, independently of rules like 'The king cannot castle out of a check' or 'The king is mate if it is in check in such a way that no move will remove it from being in check.' This contrasts with a regulative rule of chess like 'Do not relinquish control of the centre.' Even someone who disregards all of these regulative rules does not cease to play chess; he will just play chess very badly.

It can hardly be in doubt that constitutive rules are rules or norms, indeed that they make up a paradigmatic type of rule or norm. Otherwise one would have to insist that rule books do not really contain rules and that the law of contract does not really contain norms. This leaves the individualist with just one final option, namely that of insisting that, labels aside, the constitutive rules of language lack the dynamic, language-guiding oomph of genuine rules. But even this fallback position is mistaken. Among regulative rules, there are not just *prescriptive norms*, also known as 'norms of obligation' (von Wright, 1963) or 'mandatory norms' (Raz, 1978), norms of the form:

(4) *a* ought to ø in situation *S*.

Examples include: 'Promises ought to be kept,' 'Parents must look after their children,' 'Nobody should commit adultery' and 'When insulted, one should issue a challenge to duel.'

But there are also permissive norms, norms of the form:

(5) *a* is entitled to ø in situation *S*.

Examples include: 'Children under 8 may remain with their parents,' 'The chancellor can appoint a special envoy,' and 'The defendant is entitled to a legal council of his choice,' etc.

Definitions or explanations of meaning can be seen as permissive norms. For one thing, by specifying how a term is to be used, they give us a semantic entitlement, so to speak, to apply the term when certain conditions are fulfilled. That is why I can appeal to the norm in defending a particular use of an expression. If I am challenged why I call this creature a drake I can defend myself by saying that this creature is a male duck and that male ducks are called drakes. For another, the definition functions as a transformation or inference rule. It licenses transformations of empirical propositions, for example, from 'This creature is a male duck' to 'This creature is a drake' and vice versa.

Constitutive rules have a normative role. They govern descriptions, and to that extent linguistic behaviour. It is wrong to call a female duck or a male goat a drake. One of the main arguments in favour of a normativist conception of language is the fact that linguistic utterances are subject to both regulative and constitutive rules in the same way as chess moves. Someone who calls his superior an 'idiot,' or who provides lengthy explanations when he requires urgent help is imprudent, but he does not commit a linguistic mistake. By contrast, someone who violates rules of syntax, for example, by failing to conjugate properly or by violating rules of word order, or lexical rules, for example, by calling an ambiguous statement ambidextrous, commits a linguistic mistake (Schroeder, 1998, §12).

It might be thought that there are limits to the analogy with chess, on the grounds that someone who violates a constitutive rule of chess is no longer playing chess, while someone who violates the alleged constitutive rules of English, for example, is still speaking English. But this objection is dubious on both grounds. It is simply implausible to insist that a game involving a single incorrect move no longer counts as chess, just as it is implausible to insist that the medieval board game was not really chess, because the pawns could not move two squares at a time in their first move. Conversely, although someone who commits a mistake may still be said to be speaking English, the mistake itself does not count as English. It is quite common to react to a serious mistake by saying 'That's not English.' And someone who commits many such mistakes can be said not to be speaking English. What is true is that the margin of tolerance is greater and less well defined in the case of natural languages than in the case of chess, for the simple but important reason that the constitutive rules of natural languages do not form a precise and stable system.

6 Meaning, use and idiolects

That there are lexical conventions is indeed a commonplace that goes back at least to Plato and Aristotle (*Cratylos* 435d; *De Interpretatione* 16a19 and 17a1–2). The reason is precisely that what meaning a linguistic expression has is arbitrary:

- in different languages, different words are used to mean one and the same thing;
- in some languages the same linguistic form is used to mean different things ('link' in English and German);
- we could use words to mean something different from what they in fact mean.

Admittedly, there are cases of onomatopoeia. But they are rare, and even there the connection between word and meaning is far from straightforward. Expressions for animal sounds are supposed to be onomatopoetic par excellence. But German cockerels go 'kikeriki,' Swiss ones 'güggerügü,' the French, 'coquerico,' the Russians 'kukareku,' and the British 'cock-a-doodle-doo' (proof, if any were needed, that English is not a phonetic language).

But while it is *arbitrary* that we use a particular sound pattern to mean something specific, this does not entail that it is a matter of conventions.It is arbitrary what colour tag I use to label my Subject Review files, or whether before going out I start by putting on my left or my right shoe. Yet this does not indicate the existence of a convention.

In fact, some individualists try to use the arbitrariness of meaning *against* the idea that meaning is governed by conventions. Expressions do not have an intrinsic meaning. Rather, what they mean is determined by how they are used:

> ...let's not forget that a word hasn't got a meaning given to it, as it were, by a power independent of us, so that there could be a kind of scientific investigation into what the word really means. A word has the meaning someone has given to it. (BB 28)

But now the individualist argues as follows: if the meaning of an expression *e* in a language L is determined or constituted by the use speakers make of *e*, the meaning of *e* cannot at the same time prescribe a certain use to the speakers of L. Accordingly, the idea that the use of *e* is governed by lexical norms is incompatible with the idea that meaning is determined by use. It is a lapse into the Platonist picture according to which 'the meanings of words are magically independent of the speaker's intentions' (Davidson, 1990, p. 310; see 1994b, p. 12; Davidson, 1999b, pp. 41, 80; Glüer, 1999, pp. 10–11).

It would seem, however, that there is a straightforward answer to this complaint: we must distinguish between the use an individual speaker *a* makes of *e*, and the use that the linguistic community *C* makes of it. *Communal* use *constitutes* meaning, while *individual* use is *responsible* to it. By this token, the existence of lexical norms and their independence from the utterances and intentions of individual speakers is no more mysterious than the existence of legal norms and their independence from the acts and motives of individual agents.

This answer has strong roots in the hermeneutic tradition. But it can also be found in Wittgenstein's idea of a language as a communal practice. It is arguable that Wittgenstein did not regard rule-following as an *essentially* communal affair. But there is no doubt that he regarded rule-following in general and language in particular as *typically* social (see Glock, 1996, pp. 228–9).

He insists not just on the rule-governed nature of language, but also on the idea that these rules are *conventions*, that is, rules which are both arbitrary and intersubjectively shared. The grammatical rules which underlie language and logic are *arbitrary*: they do not mirror or pay heed to any putative essence or form of reality and cannot be correct or incorrect in a philosophically relevant way (PG 184–5). This position has often been regarded as conventionalist. And Wittgenstein himself had no scruples to call grammatical rules 'conventions' (*Übereinkunft, Konvention*) (PI §355; AWL pp. 89–90; PG pp. 68, 190). The term *Übereinkunft* is even more blatant in its intersubjective connotations than *Konvention*. Furthermore, in discussing the adequacy of these labels, Wittgenstein maintains that grammatical rules are akin to social norms: not their history, but their *status* and *function* are that of conventions (AWL pp.169–70). And that alone counts in philosophy, at least according to Wittgenstein.

The individualists, by contrast, challenge the very idea of a linguistic practice or language independent of individual linguistic utterances:

> ...we all talk so freely about language, or languages, that we tend to forget that there are no such things in the world; there are only

> speakers and their various written or acoustical products. (Davidson, 2001, p. 108)

The idea of a communal language governed by conventions is at best a misleading abstraction.

Both Davidson and Chomsky support this attack on the idea of a communal use/shared language by the claim that 'no two people actually speak the same language' (Davidson, 2001, p. 115). Davidson observes that 'no two speakers speak in exactly the same way,' and concludes from this that 'each speaker has, as Chomsky has maintained, a personal idiolect.' Indeed, because of the ever changing nature of prior theories, these idiolects are transitory; strictly speaking, an individual's language changes with every linguistic interaction (1994a, p. 117; see 1984, pp. 276–7; 1990, p. 311). Driven by such considerations, Davidson reaches a remarkable conclusion:

> There is no such thing as language, not if a language is anything like what many philosophers and linguists have supposed. There is therefore no such thing to be learned, mastered or born with. (1986, p. 446)

He assuages fears that this is to deny the existence of natural languages like Basque. At stake is only the idea that such a language amounts to a set of precise and specifiable rules, shared knowledge of which is both necessary and sufficient for communication between its speakers (1994b, pp. 1–2; 1994a, p. 117).

This line of reasoning is based on an untenable conception of what constitutes a (separate) language. From the fact that most people do not speak in precisely the same way it does not follow that they speak different languages. For it is not the case that every divergence in linguistic use amounts to a difference in idiolects. Someone who persistently says that it is raining whenever it is raining diverges from the common pattern of speech. But he does not speak an idiolect, he is simply an idiot. Similarly, the fact that Americans do not use the sentence 'God save the Queen!' does not show that their language is impoverished, since the rules of both idioms allow for the construction of that sentence (it shows rather that in this respect, their political system is superior to the British). Nor did the British alter their language when they stopped using 'God save the King!' in 1952.

To have a different idiolect is to differ from the linguistic community by virtue of notable phonetic deviation, or of employing a slightly different vocabulary, or the same vocabulary in a slightly different manner, or a slightly different set of syntactic rules. Davidson's view that each speaker speaks ipso facto a different idiolect suffers from an indiscriminate conception of linguistic use. It ignores that not all aspects of linguistic usage are semantically relevant. Under what circumstances individuals will use what words depends partly on what these words mean, but partly on social or psychological factors concerning the speaker, partly on features of the occasion of utterance.

In fact, Davidson himself condones a principle for the individuation of languages that is superior to identifying them with patterns of use indiscriminately conceived: 'speakers share a language if and only if they tend to use the same words to mean the same thing' (1994b, p. 3). In spite of various idiosyncrasies, speakers of English *tend* to converge in their usage as regards the meaning of words. Admittedly, there are some variations even in the semantically relevant features of language. For example, British politicians and journalists differ from the majority of Anglophones in using the noun 'sleaze' to refer to corruption, and there are people who fit Tyler Burge's famous example, in that they use 'arthritis' to refer to all rheumatic ailments, not just those afflicting the joints.

But even this kind of deviation does not bar them from speaking English. Followers of Davidson and Chomsky rightly note that 'if no two speakers ever share the very same vocabulary items or observe exactly the same rules of grammar there will be no formally precise, syntactic and semantic characterization of the language spoken by a given community' (Smith, 1998). But this only entails that linguistic communities do not share a language on the assumption that languages are individuated by such formally precise characterizations. This is clearly not how we individuate natural languages like Basque or French. Consequently there is no reason to suppose that, strictly speaking, members of a linguistic community do not share a language, provided that we use 'language' in the established way.

7 Must lexical rules be shared?

Not all arbitrary actions are conventional. Yet linguistic meaning does seem to qualify. There is no requirement that in English a certain type of canine must be denoted by the term 'dog'; 'chien' or 'Hund' might have served the same purpose. However, it is also something on which all competent speakers of English concur. Furthermore, the fact that they do so provides those eager to converse about dogs in English with a reason to do likewise.

Even Davidson acknowledges that 'society bends linguistic habits to a public norm' and that 'in practice people usually depend on a supply of words and syntactic devices which they have learned to employ in similar ways' (1984, p. 278; 1994b, p. 2, see 6). More generally, individualists can grant that we tend to speak as others do, while insisting that:

i. we do so only because we want to maximize our chance of being understood, *not* because of heeding shared lexical rules;
ii. we can communicate without conforming, as is evident from *malapropisms*.

As regards (i) Davidson holds:

> It is absurd to be obligated to a language; so far as the point of language is concerned, our only obligation, if that is the word, is to speak in such a way as to accomplish our purpose by being understood as we expect and intend. It is an accident, though a likely one, if this requires that we speak as others in our community do.

He asks rhetorically '...what magic ingredient does holding oneself responsible to the usual way of speaking add to the usual way of speaking?' and maintains that any such ingredient 'has nothing to do with meaning or successful communication,' but only with attaining some 'social status' (1994b, pp. 8–9; 1994a, p. 118; see Bilgrami, 1993, pp. 134–6).

This attack has an attractive egalitarian ring to it, but it is uncompelling. The rhetorical question is easily answered. Speakers do not just 'tend to use words to mean the same thing.' Another very real and humdrum 'ingredient' of our linguistic practice is that aberrations are corrected – persistently in the case of children and learners, subject to custom and politeness in other cases. This means that my condition (b) is satisfied. Furthermore, such corrections are standardly accepted by deviant speakers (see Dummett, 1986; Hacker, 1988) – as required by my condition (c). It is in this sense that speakers hold themselves responsible to the usual way of speaking.

Take an example uncomfortably close to home. I used to accompany some of my fumbling attempts at ordinary language philosophy by the phrase 'in common parlour.' According to Davidson's model, my colleagues should have reacted to this by silently constructing a new truth theory in which 'parlour' refers to *parlance*. In the first instance this would be a new 'passing theory' for a particular utterance of mine, and in the second instance an entirely novel 'prior theory' for Glock, one with which all future utterances of mine would be approached (see 1986, p. 443).[1] To their credit, however, my colleagues showed no such forbearance. Rather, they pointed out to me that it should be 'in common parlance' instead. I mended my ways, ruefully but gratefully. To this we can add another point: persistent mistakes are even more likely to provoke correction, whereas according to Davidson that should be treated with greater tolerance, since the deviation no longer upsets the new prior theory of hearers (Schroeder, 1998, §18). This point is connected to one made by Dummett (1994, pp. 266–7). Our insistence on linguistic conventions is not so much a sign of elitism as a long term attempt to secure facility of communication. We react more harshly to persistent linguistic offenders, because they pose a greater threat to the integrity of our shared language.

Another problem with (i) is that it suggests that the tendency of speakers to use words as others do is simply motivated by the desire to be understood, and therefore not governed by conventions (see also Bilgrami, 1993, pp. 134–5). This is a fallacy. Although, as I stressed above, conventions need not serve any prudential purposes, they can. And in such cases one can engage in a certain behaviour both because of a convention and because it is in one's interest. For example, we are well advised to drive on the left in Britain simply to avoid accidents. Yet this has no tendency to show that it is not a convention to drive on the left (see Schroeder, 1998, §15). There is no general convention to do or speak as others do. Rather, there are conventions to use particular words in specific ways. And it is prudent to speak as others do, that is, to follow the generally accepted conventions, because otherwise you are liable to be misunderstood.

As regards (ii) Davidson maintains that knowledge of shared rules is 'neither necessary nor sufficient for successful linguistic communication' (1994b, p. 2). According to a standard account inspired by Davidson's own truth-conditional theory of meaning, communication is successful only if speaker and hearer share a truth theory that provides a recursive characterization of the truth conditions of all potential utterances of a natural language L. Speaker and hearer approach each occasion of utterance armed with such a theory, and both assume that such a theory is shared by others, and assumed to be shared (1986, p. 436).

Davidson claims that this account cannot make sense of the fact that malapropisms, spoonerisms and similar infelicities are not only common, but also commonly understood (1986, pp. 440–1). In the appropriate circumstances, we have no difficulty in interpreting Mrs Malaprop's remark 'a nice derangement of epitaphs' as meaning 'a nice arrangement of epithets.' However, no antecedent truth theory for English could deliver in advance a T sentence like:

> 'This is a nice derangement of epitaphs' is true iff this is a nice arrangement of epithets

We understand Mrs. Malaprop, although her truth theory differs from ours. Therefore mutual understanding does not require that speakers and hearers share a truth theory in advance of specific speech acts:

> Each speaker may speak his different language, and this will not hinder communication as long as each hearer understands the one who speaks. ... Communication does not demand, then, that speaker and hearer mean the same thing by the same words. (1984, pp. 276–7)

What it demands instead is a 'fit between how speakers intend to be interpreted and how their interpreters understand them' (1990, p. 311).

Davidson's own explanation of how such fit is achieved runs as follows (1984, pp. 277–8; 1986, pp. 441–2). We approach each communicative situation with what he calls a 'prior theory.' But the individual members of a linguistic community differ substantially in their dispositions to verbal behaviour. As a result, speakers constantly have to react to unexpected uses by passing from their prior theory to a 'passing theory,' an ad hoc explanatory hypothesis which reinterprets any utterance in the light of the principle of charity:

> For the hearer, the prior theory expresses how he is prepared in advance to interpret an utterance of the speaker, while the passing theory is how he does interpret the utterance. For the speaker, the prior theory is what he believes the interpreter's prior theory to be, while his passing theory is the one he intends the interpreter to use (1986, p. 442)

Neither the prior theory nor the passing theory amount to a body of conventions which speaker and hearer use to communicate. The passing theory is shared between speaker and hearer, but it does not correspond to an interpreter's general linguistic competence, since 'knowing a passing theory is only knowing how to interpret a particular utterance on a particular occasion' (1986, p. 443). The prior theory, on the other hand, to some extent equips us to deal with a variety of speakers and situations. But it is neither shared between speaker and hearer, nor does it correspond to what we ordinarily think of as a natural language, since it contains many idiosyncratic features. For each speaker will constantly adapt his prior theory in the light of previous linguistic exchanges.

Davidson is right to think that malapropisms, spoonerisms and similar foibles pose a threat to the standard account. He is also right to claim that in the rough-and-tumble of everyday communication we often rely both on context-specific knowledge (e.g., concerning the speaker and the circumstances of utterance) and on 'shared general information and familiarity with non-linguistic institutions ("a way of life")' (1994b, p. 10). This is an important fact, since it casts doubt on what is known as compositionalism. Compositionalism is an idea closely associated with formal semantic theories like Davidson's own. It holds that for an understanding of sentences it is not just necessary but also sufficient to understand its components and their mode of composition. If Davidson draws the right moral from malapropisms, however, it would appear that knowledge of components and mode of composition is neither necessary nor sufficient for understanding a sentence in context. In my view he is wrong about the first point, right about the second.

Davidson is also right in holding that communication can succeed between people employing different languages. At certain stages of their development, bilingual children often employ only one language, even when spoken to in another. What is more interesting is that communication can succeed between individuals who do not share a language, even passively. In interactions with foreigners, we sometimes communicate by way of facial expressions, gestures and through intonation. The moot question is whether there is 'communication by language' without semantic conventions, as Davidson maintains (1984, p. 265).

He supports this claim by considering malapropisms. When these are understood, it clearly amounts to linguistic communication. The question is, however, whether such understanding does not rely on a background of shared conventions after all. For one thing, from the fact that people can deviate sometimes, it does not follow that they could always deviate. It is difficult to see how we might all be constantly committing malapropisms, without communication breaking down. For another, even isolated malapropisms do not support Davidson's case. We are often, albeit by no means always, capable of understanding linguistic infelicities, but only because they are recognizable and in some way, intelligible deviations from a shared norm. We would not understand Mrs. Malaprop if she said, for example, 'a nice uldmob of delgolph.'

One problem in driving home this point is that Davidson and his followers often employ standards of intelligibility that I find difficult to comprehend. Thus Davidson (1999a, pp. 108–9) assures us that 'we do not understand' sentences like:

(6) The proposition that Caesar was murdered is true if and only if Caesar was murdered

At the same time, he finds 'most of "The Jabberwocky" ... intelligible on first hearing' (1986, p. 434). I profess to the opposite inclination. To be sure, we can easily associate 'The Jabberwocky' with certain images. But we are not in a position to explain what terms like 'brillig,' 'slithy' or 'tove' mean. I am at a slightly less severe loss with the following statement by an NFL head coach that LePore and Ludwig (2002) find perfectly intelligible: 'We know that if we've got our head between our tails, we can get embarrassed very quickly.' I do get the gist of what the coach meant to say: if a team commits some kind of mistake, it will be liable to suffer an embarrassing defeat. Yet I am simply ignorant of what it means for a group of football players to have their heads between their tails. Understanding of malapropisms is pretty much guaranteed only in cases in which we are dealing with a clear variant of a single established idiom, and such understanding is clearly parasitic on the standard use.

8 Two models of successful communication

Davidson suggests that normative ingredients of language, even if real, have nothing to do with *meaning*. In fact, however, they have everything to do with literal meaning. What they do not concern is *speaker's meaning*. In his attack on conventions Davidson confuses the two (see Dummett, 1986, p. 473; Hacker, 1996a, pp. 295–9; Schroeder, 1998, §17). Thus he accuses Dummett's claim that words have a meaning independently of individual speakers of invoking 'an elitist norm by implying that people not in the right social swim don't really know what they mean' (1994b, p. 12). Proponents of linguistic conventions have never suggested that speakers might be ignorant of what *they mean* by their words, in the sense of knowing what they meant to say, but rather that they might be ignorant of *what these words mean*, or of what they actually said.

Davidson's charge is based on an explicit identification of these two notions. He writes:

> An utterance has certain truth conditions only if the speaker intends it to be interpreted as having those truth conditions.... A malapropism or slip of the tongue, if it means anything, means what its promulgator intends it to mean. (1990, p. 310)

But although Mrs. Malaprop may have intended to use a phrase meaning 'a nice arrangement of epithets,' this is not what the phrase she actually used means. Davidson's position provokes the charge that he adheres to a Humpty Dumpty theory of meaning. Davidson anticipated this charge. He wrote:

> Humpty Dumpty is out of it. He cannot mean what he says he means because he knows that 'There's glory for you' cannot be interpreted by Alice as meaning 'There's a nice knockdown argument for you'. (1986, p.440)

It is only subject to constraints of interpretability that 'what a speakers words mean' is 'what he intends them to mean.' He must have provided sufficient 'clues' to be understood (2001, pp. 13–14, 28).

Unfortunately, one cannot avoid Humpty Dumpty semantics simply by meaning to do so. No matter how many clues a speaker drops, what his utterance literally means can differ from how he intends it to be understood. This is simply part and parcel of the notion of literal meaning.

Surprisingly, Davidson professes allegiance to a Gricean distinction between speaker's meaning and literal meaning (1986, pp. 434–7). What he wants to do is to pry apart what is literal from what is conventional. His new conception of literal meaning goes by the name of 'first meaning.' The first meaning of an utterance is determined by what Davidson later called the speaker's 'semantic intention,' the intention to utter words that the hearer interprets in a certain way (1995, p. 293). It is how she intends it to be interpreted *in the first instance,* before phenomena like metaphor or pragmatic implicatures come into play. This move is problematic for two reasons. The first is that without an appeal to conventional meaning it is unclear *how to single out* such a first instance. Davidson might respond that the first instance is the one that concerns the prior theory. But if the speaker expects the hearer to interpret a metaphor as intended, this will be his prior theory, and there will simply be no mention of any literal meaning.

The second problem is that 'first meaning' is in no way equivalent to literal meaning. Imagine the following improbable scenario. Condoleezza Rice has the task of reporting to President Bush that U.S. exports are down. She knows, however, that Bush persistently confuses the terms 'export' and 'import,' and she prefers him to believe that imports are down. In that case she could say 'Our exports are down,' with the intention of inducing the belief that US imports are down, thereby giving it the first meaning that US imports are down. But of course this in no way affects the literal meaning of her utterance.

There is a wider issue here, namely a conflict between two contrasting models of successful communication. According to *communitarians*, communication is successful when speaker and hearer mean the same thing by the same words. They must assign the same meaning to the utterance in question. Thus even Davidson on occasion grants that:

> if communication succeeds, speaker and hearer must assign the same meaning to the speaker's words. (1984, p. 277)

By contrast, *individualists* insist that communication succeeds when there is a 'fit between how speakers intend to be interpreted and how their interpreters understand them' (1990, p. 311).

The Bush example shows that these two specifications do not coincide; it satisfies the individualist one while violating the communitarian one. What is more, the communitarian specification is clearly superior to the first, since it captures the idea of *mutual understanding*. The individualist specification captures only the idea of successfully pursuing 'ulterior purposes' with the help of speech acts, something which is semantically irrelevant. For there is a difference between the success of communication and the success of the intentions that the speaker pursues in communicating. In our example, Rice's utterance was successful in achieving the result she intended. But she was not successful in communicating her belief to Bush, namely to let him know that U.S. exports are down.

9 Conclusion

While Davidson sometimes purports to spell out the received notion of literal meaning on an individualist basis, other passages intimate his intention not to elucidate the concept of 'the meaning' of an expression, but to replace it by that of speaker's meaning (1994b, p. 12). The notion of literal meaning is bound up with the notion of a linguistic mistake as a deviation from a shared norm, and hence with the notion of a convention. It is important to note that these notions do not identify speaking correctly with speaking as the majority does. *Pace* Bilgrami (1993, pp. 134–6), to use *e* correctly does not mean the same as to use *e* as most people do. The latter is neither necessary nor sufficient for speaking correctly. Even in southern California one may simply say 'Hello!' and 'Good bye!' instead of 'How are you guys doing?' and 'Missing you already!,' without committing a linguistic mistake. Conversely, there is nothing incoherent in the idea that a majority should commit a linguistic mistake. For example, those who use the term 'refute' to mean 'repudiate' are now in the majority, yet they are committing a linguistic mistake.

In this particular case, there is no longer a *convention* to the effect that 'refute' means 'demonstrate to be false,' since my condition (a) is violated. Nor is there, as yet, a convention to the effect that 'refute' simply means 'repudiate,' since the older use is not subject to generally accepted sanctions. On the contrary, it would seem that even those who deviate from the old rule would still accept explanation and correction. On that condition, we can speak of a linguistic mistake even if there is no deviation from a convention in the sense of Section 3.

What this requires is only that along with a habit of employing *e* in a certain way, there is also a practice of criticizing and rectifying that employment, as required by condition (c) of my definition of conventions. In this respect there is once more an analogy between language and games. What counts as the correct way to castle in chess is not determined by what most chess players do, but by the rules. Of course, there would not be such a thing as castling, unless people engaged in a certain practice. Nevertheless, it is conceivable that a majority of chess players should castle across a threatened square. But this remains a mistake, just as long as there is a habit of ruling out such moves when recognized.

Davidson acknowledges the existence of such an idea of linguistic mistake and of correct usage. But he dismisses it as 'shallow' (1986, p. 434). Behind this verdict lies the one idea that we have yet to discuss. According to Davidson, knowledge of linguistic conventions is merely 'a practical crutch of interpretation,' one that we cannot dispense with 'in practice,' but also one we could do without 'in theory'; as a result conventions are not 'essential to verbal communication' (1984, pp. 278–80; 1994b, p. 2). What is essential, Davidson maintains, is the grasp of semantic intentions or speaker's meanings in passing theories.

It is not easy to assess this claim. It seems conceivable that humans should communicate through noises, yet aided by nothing other than intimate knowledge of each other, luck and ingenuity. The question is whether this should count as linguistic communication.

Our concept of language may not be sufficiently clear cut to provide a definite answer here. But verbal communication without conventions, and hence, as we have seen, without literal meaning, would definitely be incapable of expressing the fine grained thoughts we can express by language. It would therefore be dangerously close to the communication that animals and humans engage in with gestures and expressive behaviour, a form of communication that is definitely not linguistic. This point is connected to the unique and essential role that language plays for human beings. Language allows us to express and communicate complex thoughts even to total strangers, people who cannot be expected to be in tune with our beliefs, intentions and desires. LePore and Ludwig may not require a common language to understand what the aforementioned NFL coach meant to say, but many of us do. It is only through common mastery of a loosely defined system of conventions that such communication is possible. Human beings have more to communicate than animals, and what they communicate is not rigidly conditioned by stimuli or by circumstances. Even if it does not fall foul of the essence of linguistic communication, Davidson's scenario is oblivious to the profound anthropological function that a shared language has for our species of social primates.

According to the individualist, communication is based on constant wheeling and dealing, a matter of ingenuity, familiarity with the other communicators and sheer luck. This picture puts communication at the mercy of coincidence. It also puts it at the perhaps even more tender mercy of human intelligence. It overintellectualizes linguistic communication by demanding speakers to engage constantly in highly complex interpretations. The communitarian picture is far more realistic. To communicate, we must rely on a largely nonreflective, conventional practice. And it is this largely nonreflective base that enables us to engage in communal reflections of a very sophisticated kind.[2]

Notes

1. This view is in any event incompatible with Davidson's claim on the same page that knowing a passing theory is knowing only how to interpret a particular utterance at a particular time.
2. I would like to thank Diego Marconi for his astute comments on a first draft. I owe a more general debt of gratitude to Peter Hacker, Eike von Savigny, Severin Schroeder and Daniel Whiting for many enlightening conversations on linguistic normativity over the years.

6

Regularities, Rules, Meanings, Truth Conditions, and Epistemic Norms

Paul Horwich

1

My plan is to trace an explanatory route through the above-mentioned phenomena. I guess it's uncontroversial that they are in some way intimately related. – The question is, how. At one extreme there are philosophers who suggest that we're faced with a holistic hodge-podge: the interconnections are messy, and none of these things is objectively more fundamental than any of the rest. At the other extreme there are those who think that there is a correct order of grounding amongst them – a definite hierarchy whereby the more superficial ones reduce to the more basic ones. My own sympathies are more in line with this second point of view. However, it's not that I hold for some sort of general metaphysical reason that matters simply *must* be like that. It's rather that such a story would be particularly simple; simplicity is good if you can get it – a significant explanatory virtue; and in this case it seems to me that you *can* get it.

More specifically, what I'll be suggesting is that the basic facts here are lawlike *regularities* of word use (characterized in nonsemantic, non-normative terms); that such regularities help engender (i.e., are the primary reductive basis of) facts about which *rules* of use we are implicitly following; that these facts suffice to fix what we *mean* by our words and hence sentences; and that the meanings of our sentences (given contextual factors) determine their *truth* conditions – which we *ought* to desire to be the conditions in which they are accepted.

This picture is based on Wittgenstein's ideas about meaning and rule following. Nonetheless it is at odds with Saul Kripke's treatment of those concepts, which is billed as Wittgensteinian. And it also conflicts with the important body of work elaborated over the last thirty years by Crispin Wright – another influential exponent of Wittgenstein. So a fair part of the defense of my own position will consist in responses to their arguments.[1] At stake are both the true nature of these phenomena and the most fruitful way of reading Wittgenstein's discussion of them.

2

Normal rules are explicit instructions. Typically, we hear or read formulations of them, or think them to ourselves;[2] we understand these formulations – that is, we attach semantic contents to them; we then decide to do what is dictated by those contents; and so we conform more or less successfully. The rules of games such as chess and football are like this, as are the rules of a club, the rules of parliamentary procedure, the laws of certain countries, the rule that I impose on myself for how many times a week to go to the gym, and so on.

However, we are sometimes inclined to invoke rule following when the immediate sources of action involve no such formulation of the rule, hence no understanding and no deliberate decision to abide. For example, we speak of there being rules of English grammar. We say that the utterance, 'John seems sleeping.', goes against these rules, whereas 'John seems to be sleeping.' doesn't. But not even linguists – let alone ordinary speakers – are aware of the rules or able to state them. Nor is there reason to think that we articulate them to ourselves *unconsciously* – that they are 'written down' in some deep recess of the mind. Similarly, consider how far apart two people stand from one another when they are having a conversation. We have a sense that some distances are too close for comfort and some are too great. I've heard that feelings and customs on this matter vary somewhat from society to society. Again it is somewhat tempting to speak of rules – even in absence of articulations of them and decisions to do what they say.

Let me use the terms 'explicit' and 'implicit' to signal these two kinds of rule following. Explicit rule following is relatively familiar: – the rule is spelled out, and that formulation, in virtue of its meaning, is allowed to guide our activity. Implicit rule following is more of a theoretical posit and more murky. It's not the same as *unconscious* rule following – because that could be a matter of unknowingly, but explicitly, following a rule that is formulated in one's language of thought. The essence of implicit rule following is that no formulation of the rule is directly operative.

3

When we philosophers speak of 'the problem of rule-following' we tend to have in mind the implicit kind. Granted, one might raise the question of what are the facts in virtue of which someone *explicitly* follows the rule, R! – the rule dictating conformity with regularity R. What makes it the case that an act is, or is not, in accord with a person's explicit rule? But the initial answer to both questions is obvious: – it's the *meaning* of the rule-formulation that's responsible for these things; a written-down rule is the rule it is in virtue of its content; and if its imperatival content is to *perform an action of type A in circumstances C*, then an act (in C) is in accord with the rule just in case it is an act of type A. No doubt one then can press the issue by asking how it comes to be that a given rule-formulation means what it does. And no doubt this is a tough problem. But an answer to it will yield a full account of explicit rule following.

Implicit rule following, on the other hand, presents peculiar difficulties. For, with no formulation to help us, it is *particularly* hard to see what the facts might be in virtue of which someone would be implicitly following a given rule (rather than a different one, or no rule at all). We don't see what could possibly mark a particular act as a *violation* of the rule. Indeed, the difficulty of these questions might even persuade us to resist our initial inclination to recognize 'implicit rule following' as a genuine phenomenon.

4

However, such a retreat would be unwarranted. It's not, as some would say, that the hard questions are *bad* questions, vitiated by irrational reductionist presuppositions, and that we should be content with primitive, ungrounded, implicit rule following.[3] It's rather that, as it seems to me, an adequate explication of the notion *can* be provided.

Consider the following suggestive extract from paragraph 54 of the *Philosophical Investigations*. After first mentioning *explicitly formulated* rules of a game, Wittgenstein continues:

> – Or a rule is employed neither in the teaching nor in the game itself; nor is it set down in a list of rules. One learns the game by watching how others play. But we say that it is played according to such-and-such rules because an observer can read these rules off from the practice of the game – like a natural law governing the play. – But how

> does the observer distinguish in this case between players' mistakes and correct play? – There are characteristic signs of it in the players' behaviour. Think of the behaviour characteristic of correcting a slip of the tongue. It would be possible to recognize that someone was doing so even without knowing his language.

Along roughly these lines, I would suggest that person S implicitly follows rule R! if and only if

a. S's activity is governed by the ideal law R,
b. There is some tendency for S to correct instances of nonconformity (i.e. to react against his initial inclinations).[4]

Here, the role of condition (a) is to specify which aspect of the agent's activity fixes the *content* of his rule – that is, what makes it R!, rather than R*!, that is being followed. And the role of condition (b) is to explain why S may be regarded as *following a rule*, rather than as merely obeying a natural law (which is what, for example, the planets do in orbiting the sun). For a tendency to *self-correct* manifests sentiments of relative dissatisfaction and satisfaction, and thereby encourages talk of 'goals,' of 'what ought to be done,' and of 'violations.'[5] Let me stress – and perhaps this diverges from Wittgenstein's idea – that I am not relying on S's practice of self-correction to identify precisely which actions are to count as violations. I'm relying on it merely to motivate applying terms such as 'violation' to cases of nonconformity – cases that have been independently identified as deviations from the ideal law.[6]

5

Now one might worry about the notion of 'ideal' that is deployed here. The point of inserting it is, of course, to register that we are talking about the sort of law that has *exceptions* – for we have to accommodate the fact that rules are sometimes inadvertently disobeyed. But isn't reliance on such a notion inconsistent with our naturalist/reductionist aspirations?

I would suggest that there is no such tension, because *this* particular notion of 'ideal' is not normative, and is in no way naturalistically problematic. There is a familiar form of explanatory theory that consists in postulating a system that is governed, in so called 'ideal' conditions, by certain laws, but which is subject to a variety of factors that cause deviations from the behaviour that would 'ideally' occur. For example, only in certain 'ideal' conditions do the orbits of the planets obey Kepler's Laws; and only in certain 'ideal' circumstances do gases obey Boyle's Law. Questions of whether some such theory is plausible in a given context, and of which ideal laws and potential distorting factors should be postulated, are settled by the standard canons of scientific methodology – that is, by reference to empirical adequacy, simplicity, coherence with other successful theories, etc. Thus explanatory accounts of this kind may be thoroughly empirical and naturalistic.[7]

6

But how does my proposed account of implicit rule following differ from the one that Kripke himself considers and refutes? He assumes – rather as I am assuming – that if

> S follows rule R!

can be analyzed at all, it would have to be to something like

> S's activity would fit regularity R in ideal conditions.

However, he argues that no such account can in fact be sustained; for it isn't possible to articulate, in a satisfactory way, what is meant here by 'ideal conditions.' All attempts to do so *either* fall into circularity (e.g., 'conditions in which S would not be led to violate his rule'), *or* resort to a list of allegedly 'perfect' circumstances (e.g., 'S lives forever, has an infinitely large memory capacity, etc.') – remote circumstances whose bearing on S's activity are a matter of empirical speculation.

So what is my defence against this objection? Simply this. Kripke seems to presuppose that we are obliged to give an *a priori* specification of which conditions are ideal. And I think he's right that we can't do that. But I think he's wrong in assuming that we need to. Although my explication of rule following in terms of the notion of 'ideal law' is indeed *a priori*, one must then deploy standard scientific desiderata (including simplicity and empirical adequacy) to discover which particular combination of ideal law and potential distorting factors governs S's activity. It's an *a posteriori* matter.[8]

7

This suggestion may seem tantamount to another of the proposals that Kripke rightly dismisses: namely, that one ground

S implicitly follows the rule R!

in something along the lines of

R is the *simplest* of the *simple* regularities that roughly conform with S's activity.

Kripke's criticism is that this proposal is off target, that it misconstrues his skeptical challenge.[9] For *simplicity* is to be invoked only when we are confronted with an *epistemological* question as to which of two well understood hypotheses is true; but the present problem is to articulate what it *means* to say that that a certain rule is being followed, to articulate the *contents* of competing hypotheses of that sort.

Now, I am inclined to agree with Kripke that the account he criticizes does indeed wrongly confuse what might arguably be part of the *evidence* for a rule-following thesis with the underlying fact that would make the thesis true.[10] But my own proposal certainly respects that distinction. For what I suggested was that 'following rule, R!' consists (partly) in 'being governed by ideal law, R.' And if pressed on what this underlying fact comes to, my reply is that there is no deeper *analysis* of it, but that one can say, invoking considerations of simplicity, what kinds of data would give us reason to postulate it.

8

But this involves an idea that Crispin Wright has found objectionable – the idea (which he attributes to Chomsky) that instances of 'following rule R!' must be regarded as theoretical phenomena whose identification requires *scientific investigation*. For he claims, on the contrary, that it is via some privileged *introspective* route that we know which rules we are following. As he puts it:

> Kripke himself objects to the dispositional response that it cannot account for the normativity of understanding an expression in a particular way, intending to follow a particular rule, and so on. The reason for dissatisfaction which I have in mind, however, is not this. It is rather that Chomsky's suggestion, that the identity of followed rules is a strictly theoretical question, threatens, like the dispositional account, to make a total mystery of the phenomenon of noninferential, first-personal knowledge of past and present meanings, rules and intentions.[11]

We'll be addressing Kripke's 'normativity objection' in the next section. But, as for Wright's point, I'd say that it's far from obvious that each of us knows, independently of any sophisticated empirical research, which rules we are *implicitly* following. Conscious, deliberate, *explicit* rule following is a very different matter. There, perhaps, we *are* able to tell, simply by introspection, what we are up to. But in the *implicit* case, the most we can be expected to detect through introspection are occasional inclinations and disinclinations to do this or that particular thing. The underlying rule is available to us only via objective methods of inquiry that are equally open to everyone else.[12]

9

This brings us to what is perhaps the most widely repeated of the skeptical objections to broadly 'dispositional' accounts of rule following. It's implicit in Kripke, as Wright says, and it's echoed in a great deal of the work provoked by Kripke's discussion. The allegation is that 'naturalistic regularity' proposals such as mine could not do justice to the *normative* import of rule following, to the fact (roughly speaking) that

> S follows rule R! → S's activity *ought* (other things being equal) to be consistent with regularity R

But I would dispute this allegation. Granted, no 'ought' flows merely from S's being governed by a certain natural law – not even if it's an *ideal* law. However, once we bring in the further condition, (b), that I've argued must be satisfied in order for the operation of that law to qualify as rule following – namely, S's tendency to correct some of his own initial impulses to deviate – we *are* in a position to see why deviations are naturally considered to be 'wrong.' For the practice of occasional self-correction manifests selective *dissatisfaction*, and hence orientation towards a goal; and there is surely some positive value in the achievement of goals.

This is not to claim that rule following, under my analysis of it, *logically entails* something normative. Why should it? After all, 'S wants to do X' does not – even in the case of *explicit* desire – *logically* (or *conceptually*) necessitate 'S has reason to do X'. Yet, for all that, we do accept that S has reason to do what he wants to do; we do give normative import to a person's explicit desires. And all I am claiming for my proposed analysis is that it enables us to see why it is unsurprising for us also to give that sort of significance to *implicit* rule following.

Granted, we haven't explained that significance. We are treating the normative import of rule following – like that of explicit desire – as explanatorily fundamental. Still, we *are* able to understand why *we accept* that if

> R is an ideal law governing S's activity (in a context of occasional self-correction)

then

> S has some *reason* to conform his activity with R

and we are thereby explaining *our attribution* of normative import to rule following.[13]

10

Let me turn now to facts about *meaning*. And let me begin with a point that, I think, is not sufficiently recognized and emphasized in the post-Kripkean discussion: namely, that meaning and rule following are quite distinct phenomena and that the philosophical problems surrounding them are somewhat separate. This bears emphasis because Kripke writes as though it's all the same thing. He moves from one issue to the other in the same breath, as if there were a single topic here.

As far as *explicit* rule following is concerned, since *understanding* is essentially involved, any skeptical problems concerning meaning will indeed infect it. But such problems will equally infect all other forms of linguistic activity. There is no reason to single out rule following.

And as far as *implicit* rule following is concerned, its relationship to meaning is arguably (but not uncontroversially – see Section 14) that meaning a given thing by a word is a matter of implicitly following particular rules for its use. If that is right, then any problems surrounding rule following will be inherited by meaning. But it is worth stressing the obvious point that rules need have nothing to do with words or with language. Therefore, issues that relate specifically to meaning (e.g., the question of which particular rule is involved in engendering a given meaning) need correspond to no general problems about rule following.

Thus I don't think we can go along with Wright's remark that the rule following problem 'in its most general form' is the problem of how it is determined that a person ought (or ought not) to assert a given sentence of his language in given circumstances – that is, how it is determined that the rules governing its use dictate assent (or not) (see Wright, 2001, p. 4). For, as just noted, there are rules that do not bear upon the use of words. Moreover the role of rules in language is highly atypical. For they operate at several different levels and correspond to different kinds of normative constraint. In particular, we can distinguish (1) meaning-determining rules of use, (2) the rule instructing us to mean the same by a word as others do within our community, (3) rules of epistemic justification, and (4) the rule dictating that we try to accept only those sentences that are true. Therefore, to treat language use as a basic paradigm of rule following is to court confusion.[14] It is preferable to proceed by first trying to achieve (as in Sections 2–9) a truly general account of what it is to implicitly follow rule R!; then considering (as we will in Sections 11–14) questions of whether and how a word's meaning might derive from following rules for its use; and lastly (see Section 15) examining the important normative features of meaning, including (i) what we ought to mean, (ii) which sentences we ought to accept (in light of their meanings and the available evidence), (iii) and which we should *want* to accept (in relation to what is true).

11

If meaning a given thing by a word is a matter of following rules for its use, it must be at least sometimes a matter of *implicit* rule following. Otherwise we'd have a vicious regress. For any explicit formulation of such a rule would be in a language the understanding of which would rest upon deeper explicit rules, the understanding of which would rest upon yet deeper explicit rules, ... *ad infinitum*. To put it another way: explicit rule following is explanatorily dependent on facts of meaning; and explanatory dependence is a transitive and irreflexive relation; so it cannot be that each word's meaning what it does is explanatorily dependent on our explicit following of rules for its use.[15]

This sort of concern admittedly leaves it open that only *some* terms – our most basic ones – are understood in virtue of *implicit* rules, that certain further words are understood in virtue of *explicit* rules formulated using these most basic terms, that yet further words are understood in virtue of *explicit* rules using terms from the first two groups, and so on.

However, in the absence of any persuasive considerations to the contrary, it is not plausible that there be *two* kinds of meaning property: those partly grounded in *explicit* rule following and those grounded – very differently – entirely in *implicit* rule following (i.e., in ideal laws and self-correction). Therefore, if rule following is involved at all in the constitution of meaning, then – given the need to avoid regress – it is plausibly *always* a matter of *implicit* rule following.

Of course, one might well wonder why we should think that meaning is rule following of *any* kind. I'll come back to this issue in Section 14.

12

But if we assume for the time being that meaning *is* a matter of implicit rule following, the central questions become: (1) how are we to find out which rule (or rules) are the ones the following of which will constitute a word's having any given meaning – for example, meaning WATER, or meaning TRUE, or meaning ELECTRON; and (2) what specifically will the rules turn out to be – say, in these three particular cases?[16]

A view that has captivated many philosophers – including Kripke, I think, but he is frustratingly inexplicit – is that, for the case of predicates

> S's meaning F by word w (i.e., S's meaning by w what 'f' means in English) *is constituted by* S's following the truth-oriented rule: apply w to all and only the fs.[17]

A primary cause of the temptation to say this, I suspect, is – for example – that if S means DOG by w then (probably) S would ideally like it to transpire that he applies w to all and only the dogs. Given this meaning, that's what he will aim to do. And vice versa: – he will regulate his use of w in such a way only if he means DOG by w.

But I think we can see that this rationale is defective and that the view it is supposed to motivate is implausible.

For it is a mistake to conflate, on the one hand, *explicitly desiring that one apply w to all and only the dogs*, and, on the other hand, *implicitly following the rule: to apply w to those things*. The former (explicit) state does indeed *correlate* with meaning DOG by w. But it is explained *by* the meaning-fact rather than the other way around.[18] Moreover, this correlation gives no reason to expect that the latter (implicit) state will even so much as correlate with meaning DOG, let alone constitute it.

In addition, the implicit state clearly cannot be a purely *a priori* (conceptual) ground of 'meaning DOG.' For although we know *a priori* that our word 'dog' means DOG, we can't tell, independently of empirical research, which ideal law explains our overall use of it and hence which rules for its use we are implicitly following. In particular, we can't know *a priori* that we are *implicitly* following the rule, 'Apply "dog" to all and only dogs.'

And, as for *a posteriori* grounding (of the sort exemplified by 'salt' and 'sodium chloride'), attention to the normal methodology for settling such questions also points us away from the assumption that meaning-constituting rules are typically truth-oriented. In general, a relatively superficial property, S, is constituted by a relatively basic property, U, just in case (i) they are coextensive, and (ii) this coextensivity explains why the symptoms of S's presence are what they are. For example, the property of 'being a sample of salt' is constituted by the property of 'being a sample of sodium chloride' because the symptoms of salt (e.g., its characteristic interactions with other substances) are explained on the assumption that it coincides with sodium chloride. But the main symptoms of a word's meaning are its various ways of being used – that is, the various facts concerning the sentences containing it that are accepted and the circumstances in which this is done. So the meaning-*constituting* property will be one that explains all such acceptance phenomena; it will be the word's *basic* rule of use. So far this is *a priori*. But it's an *empirical* discovery, often quite hard to make, which particular rule is the explanatorily basic one for the use of any given word. And it does not seem likely that the rule governing our use of 'f' will typically (or even *ever*) turn out to be: apply it to all and only fs. For the *basic* rule will surely be something more readily obeyable, and will relate rather to what we do for the sake of trying to achieve that result (i.e., for the sake of trying to believe what is true). Thus, more plausibly, our use of 'electron' is guided by our acceptance of a certain physical-theory-formulation; our use of 'true' is governed by our deployment of an equivalence schema; and our basic use of 'water' stems from some rule along the lines of, 'apply it to whatever has the same underlying structure as the stuff in our rain and lakes.'[19]

13

It is widely assumed that an account of how meaning is constituted can be adequate only if one can *explain*, given the alleged meaning-constituting property, why any word that possesses that property comes to have the particular extension that it does. For example, if it is in virtue of a word's possession of property, U, that it means DOG then, according to this common assumption, it must be possible to show why it is that any word with U is true of exactly the dogs.[20] And this requirement helps motivate the presumption, just criticized, that, in general, 'f''s meaning-constituting rule will be something like: Apply w to all and only fs. For such uniformly truth-oriented rules will satisfy that requirement (as we see in note 223). But the more complex and more varied meaning-constituting rules, of the kind that I have been urging, will not. How, for example, can we get from the rule, 'Accept "That is f" when something seems red,' to 'f''s extension being the set of red things? What explanatory deduction could take us from the rule, 'Tentatively accept instances of "<p> is f ↔ p"' to 'f''s extension being the set of truths? On the other hand, we have just (in Section 12) encountered solid reasons for thinking that complex rules like these are more likely to be the right ones!

How is this dilemma to be resolved? It seems to me that we should resolve it by appreciating that the alleged adequacy condition is mistaken. One should not expect to be able to derive a word's extension directly from its meaning-constituting property. In particular, one should not expect to be able to explain, on the basis of a word's meaning-constituting rule of use, why it is true of the things it is (rather than, say, a slightly different set of things).

Why not? Because of the plausibility of the *deflationary* view of the truth-theoretic notions, 'true,' 'refers' and 'true of.' Applied to the last of these concepts – which is the pertinent one here – deflationism is the view that 'true of' is implicitly defined (relative to a prior understanding of instances of 'w means F') by our acceptance of the schema

w means F → (x)(w is true of x ↔ x is f)[21]

Assuming that this is correct, then the explanatory route leading from w's meaning-*constituting* property (e.g., that 'dog' has nonsemantic property, U) to w's truth-theoretic property (e.g., that 'dog' is true of exactly the dogs) proceeds *via* the intermediate fact that w means what it does (e.g., that 'dog' means DOG). In other words, we had no right to insist that a good analysis of (say) 'w means DOG' provide a *direct* explanation of 'w is true of the dogs' (i.e., an explanation that does not presuppose the correctness of that analysis).

Any such direct explanation would require (i) that the nonsemantic constitutor of 'w means DOG' take the form '(x)(wCx ↔ x is a dog),' and (ii) that C-ness either *constitute* 'being true of,' or otherwise bring it about. Therefore, an implication of deflationism is that we have no right to expect any such *relational* theory of meaning constitution.[22]

The deflationist-friendly alternative is not to deny that a word's meaning-constituting property fixes its extension, but to recognize that it does so only because it *first* fixes the word's meaning. Thus, 'dog''s being governed by a certain basic rule may, in virtue of the ability of that fact to explain the word's overall deployment, constitute its meaning what it does – that is, its meaning DOG. And we can then invoke the deflationist's trivial and explanatorily fundamental logical relation between a word's meaning and its extension, that is,

w means DOG $\rightarrow$ (x)(w is true of x $\leftrightarrow$ x is a dog)

to infer (by transitivity) that the word is true of exactly the dogs.[23]

14

These observations on how to determine what constitutes a given meaning property put us in a position to address the question left pending in Section 11: – 'Why suppose in the first place that meaning is a matter of rule following?'

A tempting answer might go like this. Given the above discussion, what someone means by a word stems from whatever explains his *overall* use of it; and surely a good candidate for that explainer is his following certain *basic* rules of use. Moreover, meanings have distinctive *normative* implications (e.g., if a word means DOG one had better not apply it to cats). And it may seem that this phenomenon will fall out nicely if we assume that meaning facts are intrinsically regulative.

But, on reflection, one can see that neither of these motivations is valid. Regarding the *explanatory* consideration, if my suggested analysis of rule following is roughly correct then a person's following a certain rule is grounded in *two* factors: (a) that he is governed by a certain ideal law; and (b) that this takes place within a context of occasional self-correction. But only the first of these components is needed to explain his overall use of the word. The second component is not doing any of that explanatory work, and so is not essential to the word's meaning what it does. Therefore, although we might say, speaking somewhat loosely, that someone means what he does 'in virtue of' following certain rules – for it is true that the rule following *suffices* for the meaning – a more accurate statement would be that the meaning is *constituted* merely by the ideal law of use.

And, as for the second consideration, the normative import of meaning would not in fact flow straightforwardly from the assumption that meaning is rule following. One problem is a certain mismatch between the pertinent rules and norms. As we have seen, even if 'meaning F by w' were constituted by implicitly following a certain rule, there's good reason to think that the rule would *not* be 'Apply w to all and only fs'; so it remains unclear how the obligation to apply w only to fs is to be accounted for.

Another problem is that the normative import of rule following is rather minimal; it's merely the value of satisfying the desire to conform with a certain regularity. But there is an enormous gulf between this value and the significant instrumental and epistemological value of true belief, of applying words to the members of their extensions. So, regarding meaning as a species of rule following does not, on reflection, help us to accommodate its most obvious form of normative import. (And in Section 15, we'll see that there is a perfectly good alternative way of accommodating it).

In addition, it's worth noting that, insofar as we assume (in the absence of any better account) that the contents of thoughts derive from the meanings of mental terms, and insofar as we allow that these terms may turn out to be innate and universal (i.e., symbols of Mentalese), then the idea that meaning is rule following becomes highly conjectural. For although our deployment of Mentalese symbols is likely to be law-governed, the operation of these ideal laws may well owe nothing to any practices of correction. Indeed, it may well be that 'self-correction' cannot occur at such a cognitively deep level.

This latest worry about identifying meaning with rule following goes hand in hand with another. *Correction* is an intentional state, since it involves feeling that certain responses are OK, disliking others, etc. Now, *if* these states have to be mentally *articulated*, then the meanings of the terms in which they are articulated cannot, on pain of circularity, be engendered by rule following. So perhaps one can instead take the states to be states of *implicit* desire. But it would then be necessary to give a plausible analysis of such states in nonintentional terms. However – *pace* Wittgenstein's above-quoted PI §54 – it is far from obvious that there is such a thing as 'unmistakable self-correction behaviour.' If there is, then the present objection to identifying meaning with rule following may be defective (although the others remain in force). But if not, then we have a final reason for denying that a grounding of meaning in rule following could qualify as a fully nonsemantic reduction.

The clear moral of these considerations is that a word's meaning what it does is a matter of its being governed by a certain ideal law – the one that explains its overall use. Often, such a law will operate in a context of self-correction – in which case we can suppose that a rule (dictating conformity with the law) is being implicitly followed. But the operation of a word's law of use need not be a case of 'rule following' in order for it to engender meaning.

15

What about the *normativity* of meaning? Can this be reconciled with what I have been suggesting about the reduction of meaning properties to ideal laws of use? Well, the answer is not simply 'yes' or 'no'; for it depends on what one has in mind by 'the normativity of meaning.'

If all someone has in mind is that meaning facts have certain normative *implications*, then the present story is not in jeopardy. For it easily accommodates three distinct kinds of normative implication.

The first of these concerns *truth*. Insofar as we think – as I believe we should – that true belief is desirable, then we will hold, in particular, that

> S means DOG by w → S ought to desire that (x)(S applies w to x ↔ x is a dog)[24]

And this will put us in a position to infer, given our view of how the meaning fact is constituted, that

> S's use of w is governed by ideal law R → S ought to desire that (x)(S applies w to x ↔ x is a dog)

Granted, we are not able to explain the first conditional (i.e., the value of truth) by means of the second one, since our explanatory direction is the other way around. But I see nothing untoward about this ordering. Consider, by analogy, the relationship between the fact that happiness is good and the fact that H fiber firings are good (– supposing that H fiber firings are the neurological basis of happy feelings). Presumably the former normative fact is what accounts for the latter. So it would be illegitimate to object to the constitution supposition on the grounds that it does not put us in a position to explain the value of happiness.[25]

A second sort of meaning-related norm concerns *epistemic justification*. For example: given what we mean by 'if' and 'not' we are epistemically obliged never to simultaneously accept instances of 'p', 'if p, q', and 'not q'. But – as we have just seen – one should not think that our account of meaning (or *any* account) must put us in a position to *explain* this conditional. Rather, the basic epistemically normative implications of meaning facts appear to be explanatorily fundamental.[26]

A third form of normative import that meaning possesses hinges on 'social externalism.' The ideal law of use governing a single word typically varies somewhat from person to person within a linguistic community. Perhaps most people (or the experts within the relevant area) are governed by R(w); but many are governed by slightly different laws. Nonetheless, our ordinary practice of communal meaning attribution is such that all these people are said to mean the same thing by w (– they all mean F, say). Now, these divergences in basic usage are somewhat undesirable; communication would be smoother if they didn't exist, if everyone's basic law for the use of a given word coincided with that of the majority (or that of the experts). Therefore we have the conditional

S means F by w → S *ought* to be governed by R(w).

Thus, a regularist account of meaning is able to accommodate its having yet a *third* form of normative import.

Suppose, however, that what someone has in mind by 'the normativity of meaning' is not merely that meaning facts have normative *implications*, but that they are *constitutionally* normative – that they are somehow *grounded* in normative facts. In that case my response is to concede that this idea has not been respected here – but to insist that it needn't be, since we should *not* think that 'meaning is normative' in this very strong sense.[27]

Certainly it would be a blunder to simply infer the constitutive normativity of meaning from its having normative import. Many things – for example, killing – have normative import without being constitutively normative.

Another mistake would be to try to get there by combining the assumption that meaning is rule following with the assumption that rule following is constitutively normative. For, as we argued in Section 14, it is an oversimplification to suppose that meaning is a matter of rule following. More accurately, meaning reduces to just one component of rule following, namely, being governed by a certain ideal law; and this obviously isn't constitutively normative.[28]

Finally, there is a route to the constitutive normativity of meaning that has been suggested by Paul Boghossian, amongst others. The rough idea (expressed in my own terms) is that meaning-constituting laws of use concern the circumstances in which sentences are *accepted*; but – so the argument goes – this notion of 'acceptance' is constitutively normative, since a condition of someone's possessing it is that he endorses the norm, 'We ought to accept only what is true.'

I don't have space here to do full justice to this line of thought. But, in a nutshell, my response is that there is a psychological relation we can characterize nonnormatively and nonintentionally, purely in virtue of its functional role – namely, the relation of *relying on a sentence in theoretical and practical inference*. And we can proceed to explain on the basis of that characterization why we take such 'reliance' to have a certain normative import, why we take this relation to a given sentence to be one that we *should* wish to bear to that sentence only if it is true. Such a demonstration provides reason to think that the 'reliance relation' we have functionally defined is nothing less than *acceptance* (i.e., inclusion within a person's 'belief box') and that our recognition of the truth norm applicable to it is a mere consequence of that definition rather than integral to it.[29]

Thus it seems to me that the respects in which meaning is genuinely normative – its having normative implications relative to the value of true belief, relative to the value of epistemic justification, and relative to the value of smooth communication – are phenomena that the present model is able to handle perfectly well. So there is no reason to regard meaning as *constitutively* normative.

16

I hope to have given some initial plausibility to the following explanatory picture:

i. At the base, metaphysically speaking, are facts to the effect that a given ideal law explains a given speaker's overall use of a given (mental) word.

ii. In light of practices of self-correction, his being governed by that law constitutes the fact that he is implicitly following the rule dictating conformity to it.

iii. Thus his overall use of the word stems (*a fortiori*) from his following that rule. Therefore, what he means by the word (in his idiolect) is determined by that fact. But not everything that goes into the rule following is involved. In particular, the practice of self-correction (needed to allow operation of the ideal law to qualify as a case of rule following) plays no essential role in the explanation of the word's overall use, and therefore no role in constituting its meaning. Thus the meaning reduces merely to the fact that the ideal law is explanatorily sufficient.

iv. Word meanings (so constituted) engender sentence meanings – since, when combinatorial procedure P is applied to words, <w_1, w_2, ..., w_n>, the meaning property of the resulting complex expression is constituted by the property:

> '— *is the result of applying P to words whose meanings are* <*Meaning of* w_1, *Meaning of* w_2, ..., *Meaning of* w_n>' [30]

v. Assuming context insensitivity (for the sake of simplicity), sentence meanings engender truth conditions via the explanatorily fundamental triviality: 'u means what our "p" means → (u expresses a truth ↔ p)'

vi. But true belief is valuable. In other words, one should hope that the circumstances in which someone accepts a given sentence include its just-explained truth condition. Thus, the correctness of an assertive utterance depends, in part, on the regularities of use that govern its component words.

Appendix

Wittgensteinian elements

The account developed above relies at various crucial points on insights that I take to be Wittgenstein's. And in this appendix I will briefly review the textual basis for these attributions. The result, I would say, is a solid illustration of the relevance of Wittgenstein's ideas to contemporary debates in philosophy.

(i) *Deflationism with respect to truth, reference, and extension*

The closest that Wittgenstein comes to making his semantic deflationism explicit is his remark (in PI §136) that

'p' is true = p
'p' is false = not-p

And clearly he would have subscribed also to the parallel equations

'F' is true of x = Fx

and

'n' refers to x = (n=x)

But the depth of his commitment to deflationism is manifested, not so much in such snappy formulae, as in his rejection of the Tractarian focus on meaning *qua* referent in favour of a focus on meaning *qua* use. It is only insofar as the meanings of words and sentences are constituted *non*-truth-theoretically – as he came to see they are – that the various truth-theoretic notions may then be explained trivially, as above.

(ii) *Meaning as use*

The locus classicus is PI §43:

> For a large class of cases – though not for all – in which we employ the word 'meaning' it can be defined thus: the meaning of a word is its use in the language.

But this is not something for which Wittgenstein *argues*. It is presented as evident – taken for granted from the very first paragraph of his book.

Despite the idea's supposed obviousness, it is nonetheless intended to do real work: – to demystify the 'emergence' of meaning, to remind us of what gives 'life' to otherwise 'dead' signs (*The Blue Book*, pp. 4–6), and to rectify the erroneous mentalistic view of meaning that we can easily feel tempted to espouse. So Wittgenstein should not be interpreted as imagining that word use is itself to be characterized in terms of the *intentions* or *beliefs* of speakers. For these states are themselves best understood in terms of the *meanings* of the sentences that articulate them; so such an account would not be demystifying in the above-mentioned ways. Rather, his conception of 'use' will have to be thoroughly *nonsemantic*. And his various examples of word use – 'five' (PI §1), 'slab' (PI §2), etc. – do tend to confirm that this is what he has in mind.

(iii) *Multifarious forms of word use*

Wittgenstein regarded language as a motley: the rules of word use, like the rules of a game, come in all sorts of shapes and sizes (see e.g., PI §18). So it's a confusion-breeding mistake to attempt to cram all of them into the same mold (e.g., 'Apply word w to all and only the fs') – or even the same handful of molds. In fact there's an unlimited variety of possibilities. And since many types of rule are useful to us in one way or another, many are actually deployed.

(iv) *Focus on implicit rule following, and on its elucidation in terms of 'lawlike regularity' and 'self-correction'*

Insofar as Wittgenstein holds that our language games consist in our following rules for the use of words, he cannot be thinking that these rules are followed explicitly (i.e., on the basis of formulations that must first be understood). So it isn't surprising that he both recognizes, and gives a rough account of, a different kind of rule following. To reiterate part of what I quoted and elaborated in Section 4, he distinguishes a type of game that has no explicit rules, but about which

> we say that it is played according to such and such rules because an observer can read these rules off from the practice of the game – like a natural law governing the play. – But how does the observer distinguish in this case between players' mistakes and correct play? – There are characteristic signs of it in the players' behaviour. Think of the behaviour characteristic of correcting a slip of the tongue. (PI §54)

Clearly, he supposes *this* to be the kind of rule following that is at work in our most basic *language* games.

(v) *Individualism about meaning*

Many commentators (e.g., McDowell, Kripke, Rorty, and Brandom, amongst others) have attributed to Wittgenstein the view that meaning is essentially *communal* – that a word's meaning can derive only from a practice that involves interaction between people. But I can't agree with this reading. After all, the conception of rule following suggested in PI §54 appears to be fully individualistic.

Textual support for a 'communal' interpretation is often taken to come from his denying, in PI §202, that rule following can be a 'private' matter:

> And hence also 'obeying a rule' is a practice. And to *think* one is obeying a rule is not to obey a rule. Hence it is not possible to obey a rule 'privately': otherwise thinking one was obeying a rule would be the same thing as obeying it.

But (to repeat what I said in note 4) 'not privately' doesn't mean 'not individually.' It means, rather, something along the following lines: 'not the sort of state, like an experience, that each person is peculiarly well positioned to pronounce upon – that is, not the sort of state that someone tends to be in when and only when he thinks he is in it.' The alternative to '*private* rule following' that Wittgenstein is insisting upon is 'rule following *as a practice*' – and that can perfectly well be constituted in the regular activity of a single isolated person.

But surely (some would say) he must have appreciated that the existence of *norms* of assertion depends upon language being a social phenomenon; surely he saw that only then would there be a criterion of *correct* use – namely, usage sanctioned by others. Not at all! This suggested criterion is highly implausible – as is the (inflationary!) assumption that the possibility of correctness requires some *criterion* of correctness. Unsurprisingly, the idea is nowhere to be found (or even hinted at) in the *Philosophical Investigations*.

(vi) *Nonnormative constitution of meaning*

I argued in Section 15 that although meaning has normative *import* it isn't *constitutively* normative. In particular, the fact that a word has the meaning it has does not reduce to a fact about how one *ought* to use it. In accord with this position, Wittgenstein's various characterizations of 'meaning-engendering word use' never suggest that he is taking this to be a normative idea. None of his illustrations rely on normative vocabulary (– e.g., our use of 'five' when buying apples, and the way 'slab' is used within the builders' simple language game). – And nor do his alternative general articulations of the idea: – for example, 'ways that one operates with words' (PI §1), 'functioning of the words' (PI §5), 'how one acts upon a word' (PI §6), 'its application' (PI §11), etc. It's how a word is used that he takes to provide its distinctive meaning.

(vii) *Antitheoretical metaphilosophy*

It will seem to many that there is a problematic tension between the allegedly Wittgensteinian picture proposed in this essay and Wittgenstein's official *meta*philosophy (PI §§89–133), which prohibits philosophical theorizing. Indeed, quite a few Wittgensteinians have responded to this supposed tension by maintaining that the metaphilosophy should not be taken seriously.[31] And many others have gone the other way, arguing that Wittgenstein himself – given his metaphilosophy – couldn't possibly have had anything as theoretical in mind as the line of thought that I have been developing.[32]

But my own view is that no such tension really exists. The picture I have sketched is not 'theoretical' in *Wittgenstein's* sense of the term, but is nothing more than an assemblage of potentially obvious points. Of course, what is *potentially* obvious may in fact fail to be recognized by someone who is looking in the wrong direction – distracted by confusion, by overstretched analogies. But the hard work needed to expose and rectify such errors is not the same as *theory* construction properly so called.[33]

Notes

1. See Kripke 1982. And see Wright's collection of papers, (2001) – especially his (1989). See also a more recent essay of Wright's – 'What is Wittgenstein's point in the rule-following discussion?' – presented to the NYU Language and Mind Seminar in April 2002, but as yet unpublished.
2. Note that even when someone's rule is not overtly articulated – even when there is merely the active intention to conform with a given regularity – this state may nonetheless be *explicit*, that is, it may involve a mental *formulation*, a mental 'conveyor' of its propositional content.
3. Antireductionists about rule following include Kripke, McDowell, Wright, and Brandom. According to these philosophers (but in contrast with what I am about to suggest) Wittgenstein too would have been unsympathetic to any reductionist account.
4. With Wittgenstein, I see no need to insist that the correction practices involved in implicit rule following must stem from *other* people, and therefore no reason to reject the possibility of *individualistic* rule following. Granted, he says (PI §202): 'And hence also "obeying a rule" is a practice. And to *think* one is obeying a rule is not to obey a rule. Hence it is not possible to obey a rule "privately": otherwise thinking one was obeying a rule would be the same thing as obeying it.' But what he means here is that someone's following a rule requires more than her subjective impression of doing so: an objective regularity ('practice') must be there. He is not backing away from his earlier (above-quoted) remark, which would suggest that someone's following a given rule, R!, might be an *intrinsic* property of that individual.
5. There will be further discussion of the normative import of rule following in Section 9.
6. We can't simply identify cases in which S deviates from his rule with cases in which he corrects himself (or has a disposition to do so) – because he may well fail to notice certain cases of noncompliance. Thus S's rule cannot be straightforwardly read off his practice of self-correction. Nonetheless, that practice is an important part of the empirical evidence that can help us (in the way indicated in Section 5) to reach plausible conclusions as to which combination of ideal laws and occasional distorting factors are influencing S's activity.
7. It might be objected that there can be no determinate fact as to *the* (unique) ideal law governing a given phenomenon. For *various* idealizations – various

suppositions of the form, 'such and such complicating factors are absent' – are always possible; and so it is always up to us, in light of the purposes at hand, to *stipulate* which such simplifying suppositions we are making on a given occasion. But what I have in mind here by '*the ideal* law' is the law that obtains in the *scientifically best* idealization – the one that best combines simplicity, empirical adequacy, and the satisfaction of our other methodological desiderata. Such 'ideal laws' are also known as 'ceteris paribus laws.' Outside fundamental physics, nearly all scientific laws are of this kind. Granted, there may *sometimes* be more than one 'equally best' idealization. And in such a case – provided the 'correction condition (b)' on implicit rule following is satisfied – we would have to say that there is a certain degree of *indeterminacy* as to which rule is being implicitly followed. In particular there may well turn out to be such indeterminacy as to which particular rules implicitly govern our use of certain words. However – as we shall see in Section 13, and contrary to what one might be tempted to think – this would *not* engender any indeterminacy in the *meanings* or *extensions* of those words. – I would like to thank Paul Boghossian and Allan Gibbard for pressing me on the points in this footnote.

8. One might suspect that the problem of specifying 'ideal conditions' cannot be so easily disposed of. For in the case of *certain* rules – for example, 'Apply word w to all and only electrons' and 'Accept "x plus y = z" *iff* x plus y = z' – it is plausible that the difficulty in coming up with conditions necessary and sufficient for our conformity to reflect obedience to them is not merely that no such (nonsemantically specifiable) conditions can be discovered *a priori*, but that no such conditions *exist*! I would concede this point. But I'd argue that we can happily embrace what it implies: namely, that *not* all rules of the particular form – 'Apply w to all and only the fs' – *can* be *implicitly* followed. This will not in turn imply that certain predicate meanings are irreducible. For, as we'll see in Section 12, it's a mistake to assume that the implicit rule following that engenders 'meaning F' is always of the truth-oriented rule, 'Apply w to all and *only fs..*' Rather, an unprejudiced empirical investigation of which factors are governing overall word usage will turn up ideal laws (hence, rules) of a variety of very different forms.
9. See his (1982), p. 38.
10. Kripke remarks that the operative confusion results perhaps from 'the influence of too much philosophy of science.' It's not clear exactly what he has in mind here. But maybe he is expressing disagreement with the Ramsey-Lewis account of what it is to be 'a basic law of nature' (roughly, 'One of the axioms in the simplest of the simple axiomatizations whose theorems characterize all (and only) the facts'). And maybe he thinks that this piece of philosophy of science is an especially prominent example of confusing evidential conditions with constituting conditions. Of course, the evidential import of simplicity is no bar to the existence of concepts that are best explicated in terms of the notion of simplicity, and hence no bar to the possibility of hypotheses whose contents are to be characterized by reference to that notion. So in giving an analysis of rule following in terms of simplicity someone needn't be failing to see that a metaphysical (constitution) thesis is called for. Therefore, Kripke's criticism is best articulated as follows: – that

the proposed analysis is *incorrect*, and that it's air of plausibility stems from moving too quickly from a theory of evidence to a theory of constitution.

11. From Wright (1989), reprinted in Wright 2001 – see especially pp. 174–7. Kripke sketches a similar argument on pp. 39–40 (including footnote 25) of (1982).
12. Both Wright and Kripke also maintain, in the same vein, that a word's *meaning* what it does cannot be constituted by *theoretical* facts (for we surely know *directly* what we mean). However, the only fact as to what (say) 'plus' means that ordinary speakers *explicitly* know is *that it means PLUS* and *that it is true of the plus triples* – and, granted, that knowledge is immediate and trivial. But such facts, although perhaps *a priori*, are nonetheless contingent, and may perfectly well be empirically reducible to facts that are not 'directly' accessible. Compare the relationship between 'Our rain, rivers, lakes and seas consist mainly of water' (which is more or less *a priori*) and 'Our rain, etc. consists mainly of H2O' (which is a scientific discovery).
13. It may be, strictly speaking, that one has reason to *conform* with one's rule only if it isn't *unreasonable* to be *following* it. But presumably this insertion of normative content into the antecedent of the conditional at issue could only make it *easier* for a regularist to accommodate its normative consequent.
14. For example, consider the following passage: 'Is not the image of the syntactic-semantic engine simply a graphic expression of the rule-governed nature of linguistic practice? And if we reject it, are we not, in effect, rejecting the idea that proper linguistic practice is rule-governed – and thereby rejecting the notion of strictly *correct* linguistic practice, and with it the notion of meaning?' (Wright, 2001, p. 180) – From a perspective in which the various above-mentioned rules and norms are kept distinct from one another, there will be *little* temptation to think that if word use were not a matter of following rules then the notions of correctness (i.e., truth) and meaning could not enter the picture.
15. I am taking the view that a person's *explicitly intending* to conform to R depends on there being a certain combination of mental elements (within his medium of thought) that possesses the meaning, CONFORM WITH R. So – given the present hypothesis that meaning is rule following – this content would have to derive from those elements themselves being rule-governed.
16. I employ the convention of writing expressions in small capital letters in order to refer to their meanings. Thus 'WATER' is to name the meaning of our (English) word, 'water', etc.
17. Arguably, the rule here should instead be something along the lines of: 'Apply w, when queried (or when considering the matter), to any f and only to an f.' But, for ease of exposition, I'll continue to work with the simpler formulation.
18. Here's the explanation. Suppose S understands '#'. In that case he (probably) explicitly desires the truth of the proposition that he would express by '(x) [I apply "#" to x ↔ #(x)]'. (I'm pretending that 'I' and 'apply' are in S's language). Now assume that '#' means DOG – that is, that '#' means the same as 'dog.' Then, the content of his desire is that he apply '#' to all and only dogs.
19. For further discussion of the question of how to determine which particular nonsemantic property of a word provides it with meaning, see my (2005), chapter 2.

20. Closely related to this 'explanation requirement' is the requirement (see Kripke, 1982, p. 26) that one be able to 'read off' from each alleged meaning-constituting property what the extension would be of any word possessing it. Insofar as this 'reading off' would be made possible by a general theory that reduces 'meaning F' to 'bearing naturalistic relation R to fs,' then these two requirements will coincide. (For one might explain, by reference to that general theory, why, for example, w means DOG if and only if w bears R to the dogs). But suppose we come to suspect that meaning-constituting properties are all of a certain kind – that they are all Π-properties – where these have no such relational form. In that case, if we are given the information that w has the very Π-property that is possessed by our predicate 'f,' we could nonetheless infer ('read off') that w is true of all and only fs. But we would not have explained *why it is* that if a word has the same Π-property as our 'f' then it is true of fs. My hunch is that the more fundamental intuition here is the 'explanation requirement,' and that the 'reading off requirement' is an imperfect expression of that intuition.
21. For the case in favour of deflationism, see my (1998a).
22. Note that if we were to take the view, criticized above, that 'S's w means F' reduces to 'S follows 'Apply w to all and only fs,' then we would have a relational theory, and so the requirement in question could be satisfied. For, supposing that *S's w is true of x* is engendered by $(\exists y)$*[S follows 'Apply w to all and only ys' and y(x)]*, we can see how the proposed analysis of 'S's w means F' will entail, and hence *directly* explain, why S's w is true of all and only fs.
23. For further discussion of the explanation and reading-off requirements, and their affiliation with inflationism about truth, see Horwich, 2005, chapter 3. See also my 'Kripke's Paradox of Meaning.' Note that it is perhaps consistent with deflationism for us to discover (via the proper methodology) which properties constitute the meanings of a variety of words; to notice that, no matter what w means, the nonsemantic ground of 'w means F' always entails '(x)(wCx $\leftrightarrow$ fx)' (where 'C' is constant); and *then*, in light of that observation, to reduce 'w is true of x' to 'wCx.' What deflationism clearly precludes is the *advance assumption* that any adequate reductive account of meaning *must* have that character. Thus it is one thing for us to find to our surprise, in light of the meaning-constituting proposals that we have discovered to be correct, that 'explanations of extensions' and 'reading off' are possible. But it is another thing to impose that possibility as a requirement in order to filter out 'bad' proposals. That would be tantamount to our assuming *a priori* that 'w is true of x' had some reductive analysis or other.
24. Nothing hinges here on whether this is really the best precise way of articulating the idea that truth is valuable. For discussion see Horwich, 2006.
25. I owe this example to Anna-Sara Malmgren. – Note that when a constitution thesis is advanced as a *conceptual analysis* then the situation is very different. For when predicate 'S' merely rearticulates the meaning of 'U,' then we can indeed insist, if we accept 'S is good', that our acceptance of it be explained by our acceptance of 'U is good'. Thus, it was incumbent on me (in Section 9) to explain our attribution of normative import to rule following by first showing that the suggested conceptual analysans of rule

following would be given that normative import. But my proposed reductions of meaning properties to specific basic laws of use are to be established *empirically*. If we focus instead on what we are taking to suffice, *a priori*, for w's meaning F – that is, w's being governed by the very law that governs our use of 'f' – then we *can* thereby understand the normative import of meaning: we can see why it is that if S's word 'g' means F, then S should desire to apply 'g' to all and only fs. For suppose S's law for 'g' is the same as his law for 'f'. Then he will accept 'The fs are the gs'. Therefore, since his desire box should contain 'I apply 'g' to all and only gs,' it *should* also contain (via practical inference) 'I apply 'g' to all and only fs'.

26. For supporting arguments see Horwich, 2008.
27. Philosophers who have urged the intrinsically normative nature of meaning include Paul Boghossian (2003b), Robert Brandom (1994), Donald Davidson (1984), Alan Gibbard (1994), Mark Lance and John Hawthorne (1997), and John McDowell (1984).
28. Moreover, the same can be said of the other component as well: namely, the practice of occasional self-correction. This, I have argued, does have normative import; but is, in itself, a purely factual phenomenon.
29. See Boghossian, 2003b; see also Shah and Velleman, 2005. For further critical discussion of this line of thought, see Horwich, 2005, chapter 5. For a sketch of how 'acceptance' might be functionally defined in purely factual, nonsemantic terms, see Horwich, 1998b, pp. 94–6.
30. For details of this approach to compositionality, see Horwich, 2005, chapter 8.
31. See, for example, Rorty, 2009, which argues in favour of what he calls a 'pragmatic Wittgensteinianism' that jettisons the metaphilosophical pronouncements. My 'Rorty's Wittgenstein' (in the same volume) offers a critical response.
32. See, for example, John McDowell's (1992).
33. The first draft of this chapter was written for a workshop on Knowledge, Rules, and Reason, held on 4–6 November 2005 at the Wye River Conference Center, under the auspices of the Department of Philosophy of Johns Hopkins University. A subsequent draft was presented at a conference entitled, 'Is There Anything Wrong with Wittgenstein?,' arranged by the editors of this volume, and held at the University of Reggio Emilia in June 2006. I would like to thank all those who participated in these discussions for their helpful questions and comments. In addition, since the present chapter – except for the Appendix – appeared first in Coliva (ed.) 2009, I must thank those publishers for their permission to reprint it here.

7
Wittgenstein and Necessary Facts

Diego Marconi

1 Introduction

A long philosophical tradition in modern philosophy, stemming from Kant, holds that necessary features of the world are a by-product of our conceptual scheme (or schemes): we regard as necessary what we could not experience, or conceive, or linguistically describe as being otherwise. Necessity does not inhere in the things themselves: it is projected onto them by us. Against this tradition, nowadays many believe that things have essential (hence necessary) properties independently of how we experience, conceptualize, or describe them, and that certain facts – certain ways things stand – are necessary, that is, common to all possible worlds in which their constituents exist. Kripke's arguments for some truths being both necessary and *a posteriori* are commonly regarded as having established the latter thesis.

Wittgenstein, both early and late, consistently argued that there are no necessary facts. In addition, the later Wittgenstein – as opposed to the author of *Tractatus* – insisted that necessity stems from our 'form of representation': in this respect, he fully belongs to the Kantian tradition. This may be one of the reasons his philosophical influence has been fading in the last decades. He has been (correctly) seen as an opponent of the very idea of *direct* metaphysics, that is, of a philosophical theory that aims at outlining the structure of reality independently of our description, or conceptual representation of it. This paper intends to reconsider the issue of whether direct metaphysics is a viable philosophical enterprise, from the vantage point of the question whether there are necessary facts – as the Kripkeans claim and Wittgenstein denied.

First, I will examine an argument against necessary facts that can be reconstructed from *Tractatus logico-philosophicus* (§3). I will show that

Wittgenstein's argument, as it stands, heavily depends on Tractarian assumptions that few among today's philosophers would accept (§4). However, there is a class of simpler arguments that, though sharing the basic inspiration of Wittgenstein's argument, are not equally dependent on controversial Tractarian tenets (§4). Like Wittgenstein's, arguments in this class also conclude that no proposition can be both necessary and *a posteriori*.

I then show that there are three strategies to undermine arguments in this class (which I collectively label 'the Simple Argument against necessary facts'). The first strategy consists in denying that whatever is conceivable is possible. The second (Kripke's strategy) consists in denying that we can really conceive certain facts to be otherwise (e.g., that salt is not NaCl). The third (Chalmers's) consists in showing that the Simple Argument is fallacious by equivocation. I then examine each strategy in turn. Kripke's strategy (§6) is seen to be liable to two serious objections. Chalmers's strategy (§8) is regarded as sharing some of the weaknesses of Kripke's, and, besides, relying on rather obscure notions of conceivability. The most promising strategy tries to show that the inference from conceivability to possibility is invalid (§8). However, in its strongest version it relies on a notion of conceivability on which rejection of the conceivability → possibility inference is equivalent to the assumption that there may be necessary facts. Hence, the best strategy against the Simple Argument only succeeds in showing that *if* the notion of a necessary fact is viable then the Simple Argument fails, while it does not succeed in rejecting the Argument on independent grounds, that is, without assuming that there may be necessary facts. I conclude (§9) that the issue of necessary facts – the issue of *a posteriori* necessity – should be regarded as still undecided, as far as argument goes.

2 Kripke vs. the Kantian tradition on metaphysics

Kripke's 'discovery' of the necessary *a posteriori* has been called 'one of the great philosophical achievements of the twentieth century', that has 'transformed the philosophical landscape, recalibrated our sense of what is possible, and reshaped our understanding of our own philosophical past' (Soames, 2003, Vol. 2, p. 456). I share, to a large extent, this assessment of the import of Kripke's well-argued claim that there are propositions that (if true, as they are usually taken to be) are both necessary and *a posteriori*. I believe that one way it 'transformed philosophical landscape' is by legitimizing straightforwardly metaphysical talk of necessity, in direct opposition to a theoretically powerful and influential line of 'critique of metaphysics' that had intended to purify talk of necessity of its traditional, metaphysical implications.[1]

For a significant, perhaps dominant trend in modern philosophy, stretching from Kant to the later Wittgenstein, allegedly necessary features of the world are really features of our experience of the world, or of language, or of our system of representation. These philosophers do not *stricto sensu* deny that there are necessary features of the world: for example, they do not deny that, necessarily, every thing is identical with itself. They insist, however, that the necessity of such features comes from how we experience, or linguistically describe, or conceptualize the world, not from the world itself. Whenever we attribute a necessary feature to the world we are really projecting onto it some feature of the conceptual scheme we are adopting.[2] It is not just that what we take to be world knowledge is really derived, in its entirety, from the properties of our conceptual scheme (so that it is entirely *a priori* epistemologically): the alleged necessary features themselves depend on our conceptual scheme. Such features are necessary *with respect to* our adopted conceptual scheme and *only because* we are adopting it: given a different conceptual scheme, they might turn out not to be necessary (or even not to be features of the world at all). Therefore, (traditional) metaphysics goes wrong not so much in regarding certain propositions as necessarily true, but in taking necessity to be a property of states of affairs 'as they are in themselves', independently of all conceptualization. For this Kantian trend, there is no *de re* necessity.

Though the point is one, different philosophers put it in substantially different terms. For Kant, 'The categories of modality ... only express the relation of the concept to the faculty of knowledge';[3] in other words, saying that something is possible, or actual, or necessary amounts to saying that its concept bears a certain relation to the faculty of knowledge. In particular, in the case of necessity Kant says that 'That which in its connection with the actual is determined in accordance with universal conditions of experience, is (that is, exists as) *necessary*.'[4] Thus saying that a concept F is of something necessary amounts to saying that, given the conditions of experience, F is instantiated no matter what ('its connection with the actual is determined in accordance with universal conditions of experience'). Hence, that something is necessary depends on the structure of experience.

For Carnap, necessity is relative to a language. A sentence of a language expresses necessity if and only if it is true and its truth value is determined by the semantic rules of the language (Carnap, 1947, II). If a language L includes among its semantic rules (perhaps as a meaning postulate) that salt is NaCl, then it is necessary that salt = NaCl relative to L; otherwise, it isn't (as far as L is concerned). Moreover, Carnap thinks that there can be no *theoretical* reasons to choose one particular language or conceptual framework over another. Our reasons for sticking to, say, the framework of modern physics are pragmatic reasons; different and better, though no less pragmatic reasons might persuade us to adopt a different framework instead. Hence, not only it doesn't make sense to say that a proposition is necessary 'absolutely'; we cannot even say that it is (so to speak) *eminently* necessary, as it turns out to be necessary relatively to the *right* conceptual framework. At most, we can say that there are good pragmatic reasons to adopt a framework on which the proposition is necessary.

In the *Tractatus*, 'the only necessity that exists is *logical* necessity' (6.37); the only necessary relations are truth conditional relations among propositions. Allegedly necessary (pseudo) facts such as 1 being a number or there being only one 0 are not pictured by meaningful propositions; what they try to say can only be shown, and is in fact shown by how the symbols '1' and '0' function in language (4.1272). This generalizes to all necessary features of the world, some of which are seemingly described by *Tractatus* sentences: for example, that things are possible constituents of states of affairs (which the *Tractatus* says is 'essential' to them, 2.011), that a spatial object is situated in infinite space (2.0131), etc. None of such features can be meaningfully described; indeed, *Tractatus* sentences attempting at describing them are meaningless. Necessary features of the world can only be grasped by understanding how symbols are used in language. This, in itself, does not make them dependent on features of language more than it makes such features of language dependent on features of the world. Thus, even though according to *Tractatus* there are no necessary truths (as there are no meaningful propositions that are necessarily true) this is not to say that there are no necessary features of the world. If metaphysics is seen as an attempt at describing the world's necessary features then it is bound to fail, for it is bound to issue in statements that are meaningless; even *asking* a metaphysical question yields nonsensical formulations. Nevertheless, what metaphysics is vainly striving to describe is, somehow, really there. In this respect the *Tractatus* does not fully belong with the Kantian tradition: though necessity is only manifested through the properties of language, it inheres to the world itself no less than to language.

Not quite so with the later Wittgenstein. In his later writings, Wittgenstein consistently interprets alleged necessary truths as rules of grammar – that is, 'deep' linguistic norms – in disguise. 'Whenever we say that something must be the case, we are using a norm of expression' (AWL, p. 16); '*Essence* is expressed by grammar' (PI §371); 'Do not say "one cannot"', but say instead: 'it doesn't exist in this game' (Z §134). Whenever we feel we are facing objective necessity or essential properties, we are really dealing with our form of representation, as Wittgenstein more profusely adumbrates in a rather obscure text of 1936:

> By an optical illusion, we seem to be seeing in the thing's inside what is inscribed in our eyeglasses. 'But *this* is how it is –' I say to myself over and over again. I feel as though, if only I could fix my gaze absolutely sharply on this fact, get it in focus, I must grasp the essence of the matter. For it seems to lie in the matter's inside. As soon as this optical illusion is removed, we can see language simply as it is. However, the expression of this illusion is the metaphysical use of our words. For we predicate of the matter what belongs to the form of representation. We mistake the possibility of comparison [scil. with the matter], by which we are impressed , for the perception of a situation of the utmost generality.[5] (MS 142, p.115; = TS 220,[6] p. 88)

We see facts as superfacts (i.e., *necessary* facts) for we project features of our form of representation onto facts. This is what metaphysics is all about: metaphysics consists in taking features of our form of representation to be features of the world. In this sense 'it obliterates the distinction between factual and conceptual investigations' (Z §458).

As is well known, for the later Wittgenstein the rules of grammar are not grounded in the nature of things. To the extent that they can be said to have grounds, such grounds lie in our needs: 'To the *depth* that we see in the essence there corresponds the *deep* need for the convention' (RFM I §74b); 'We may be irresistibly attracted or repelled by a notation' (BB p. 57). Our linguistic/conceptual framework is never forced upon us by the nature of things, even though somewhat different facts of nature might make a different framework intelligible for us (PI II xii). At most, what we take as necessity can be said to be grounded in our (contingent) form of life.

Kripke broke with this tradition. For him, there are necessary features of the world that do not depend on our description of the world, but on the nature of things. As such, they need to be discovered by investigating nature; they cannot be drawn from transcendental analysis of experience, or from the analysis of the meaning of words, or from reflection on our conceptual framework. They are, in this sense, *a posteriori*. More precisely, *that* the world has such features – for example, that salt is NaCl – can only be known *a posteriori*, if at all; whereas we know *a priori*, thanks to philosophical argument and, perhaps, intuition, that such features are necessary features (that salt is necessarily NaCl if it is NaCl at all). Thus, in addition to necessary truths that are discovered by *a priori* reflection (be it philosophical or mathematical) to be both true and necessary, such as analytical truths or mathematical truths, there are necessary truths that are not discovered to be true by *a priori* reflection, though they are determined to be necessary by philosophy.

If, like Wittgenstein in the *Tractatus*, we call 'fact' any arrangement of things that is actually the case, then the claim that there are necessary features of the world – that there are ways things stands that are necessary – can be stated as the claim that *there are necessary facts*. For the Kantian tradition, there are no *inherently* necessary facts, so to speak: as features of things depend on our conceptualization (description, etc.) of them, whether or not they are necessary also depends on how we conceptualize them. Consequently, every necessary feature of the world can be known *a priori*, as it derives from the structure of our encounter with the world in experience or in language (it is, indeed, a feature of such structure). By contrast, for Kripke some necessary facts cannot be known *a priori* for it cannot be known *a priori* that there *are* such facts. Once it is established that they are facts, it is indeed *a priori* that they are necessary: once we know the world has such and such features, it follows *a priori* that they are necessary features ('$p \rightarrow$ Nec p' is *a priori*). However, that the world does have such features cannot be anticipated *a priori*: it can only be discovered through empirical research. Within the Kantian tradition, if the world has necessary feature F – for example, if substance is necessarily permanent through change[7] – we know *a priori* that it has F: we do not have to wait for experience to determine that substance is preserved (in its *quantum*) in spite of changes it may undergo. This is because it is *a priori*, as deriving from the structure of experience, not just that $p \rightarrow$ Nec p, but that p.

What difference does this make for metaphysics? According to some, a vast difference. For in the Kantian tradition, when we are speaking of necessity we are speaking of ourselves, not of the world. Hence, the Kantian tradition is deemed responsible for taking all the excitement out of metaphysics (which may be just what it intended to do): for such excitement is seen as attaching to the possibility of getting hold of the objective structure of the world, not of exploring a conceptual apparatus that could, in principle, be radically different from what it is.[8] Kripke (so it is claimed) gave us back genuine metaphysical necessity that the

Kantian tradition had misunderstood as the structure of subjectivity or – even worse, in the positivists' case – as arbitrary convention.

This is, I believe, one important reason for the decline of Wittgenstein's influence within analytic philosophy in the last three decades. The later Wittgenstein, as a deeply original but essentially faithful innovator within the Kantian tradition, constantly interpreted allegedly metaphysical knowledge as grammar in misleading disguise. Like Kant, he saw necessity as stemming from *a priori* constraints we impose on reality; like Carnap, he thought that such constraints had no theoretical foundations. Within the post-Kripkean *Zeitgeist*, nothing could be more obnoxious. Nevertheless, Wittgenstein's views were supported by arguments that – in one version or another – are still with us.

3 Wittgenstein's argument against necessary facts

Wittgenstein's *Tractatus* contains a powerful argument against necessary facts. Though – as so often – the argument is not explicit, it can easily be reconstructed from Tractarian materials. It goes as follows:

1. Suppose there are necessary facts, and let the fact that A be such a fact.
2. As A is a fact, there is a proposition that pictures it; let it be P.
3. As P pictures a necessary fact, its negation, ~P, says that a necessarily obtaining state of affairs does not obtain, that is, it states an impossibility.
4. Whatever is impossible is unthinkable.
5. Hence, what ~P says is unthinkable.
6. Consequently, ~P is not a genuine proposition.
7. Therefore, P itself is not a genuine proposition.
8. Hence, P does not picture any fact.
9. But P had been said to be a picture of a fact, A.
10. Hence, either A is not a fact or it is not necessary.

Comments:

(2) If there were facts that cannot be pictured, it would not be true that the world is completely described if all true elementary propositions are listed (4.26; see also 3.01).

(4) 'What is thinkable is possible too' (3.02). (4) follows by contraposition.

(6) Suppose ~P were a genuine proposition; then, it would be a picture of some state of affairs or other. In that case, what ~P says would not be unthinkable. Hence, ~P is not a genuine proposition.
(7) Could P be a genuine proposition though ~P isn't? Aside from being ruled out by Tractatus (4.0641, 4.0621c, 5.2341c, etc.), this would be odd on any theory of language.

As we shall see in a moment, the most controversial steps in Wittgenstein's argument against necessary facts are (4) and (6). In fact, theses equivalent to or identical with (4) and (6) will most often be challenged by friends of necessary facts, such as Kripke and Putnam. On the other hand, both (4) and (6) are deeply rooted in the early Wittgenstein's conception of language. Ultimately, (4) derives from two premises: that thoughts are pictures, and that pictures are of inherently possible situations. The second premise is a cornerstone of the *Tractatus'* picture theory: it is connected with the basic intuition that genuine propositions are capable of being both true and false. The first premise follows from Wittgenstein's Fregean view that meaningful (i.e., genuine) propositions express thoughts (3.1), given the picture theory of language.

(6), in turn, derives from Wittgenstein's somewhat idiosyncratic notion of thought as the *abstractum* of the notion of picture. Particular pictures are characterized by specific pictorial forms: thus there are spatial pictures, chromatic pictures, sound pictures, etc. However, *any* picture of a situation S must consist of entities, standing for elements of S, whose relations are regarded as going proxy for the relations among the correlated elements of S; that is, any picture must be a structure that is appropriately correlated with the situation it is a picture of. Such an *abstractum* of pictorial form, which every picture, whatever its specific form, must share with the depicted situation, Wittgenstein calls 'logical form'. A picture whose pictorial form is *just* logical form is a *thought* (2.181, 3).[9] Thus every picture of S involves a thought of S. Therefore, nothing unthinkable can be pictured, as there is no picture of it that does not involve a thought of it.

Both the notion of logical form and the picture theory are crucial to the *Tractatus'* framework: Wittgenstein's argument against necessary facts is deeply entrenched in his early semantics-*cum*-ontology. One could not countenance necessary facts without giving up most of the *Tractatus*. This indicates an easy way to get rid of Wittgenstein's argument: as the picture theory is deeply unsatisfactory anyway and the notion of logical form is obscure to say the least, let us forget about the whole *Tractatus* business, including its veto against necessary facts. As we shall see, things aren't quite so simple.

4 Getting rid of Wittgenstein's argument

Let us look at the two most natural ways of rejecting Wittgenstein's argument. The first consists in rejecting (4), that is, in holding that at least some impossibilities are conceivable: contrary to *Tractatus* 3.02, the inference from 'conceivable' to 'possible' is dubious. However, as we just saw, on Wittgenstein's notion of conceivability or thinkability the inference, far from being dubious, is quite straightforward. Thus, rejecting 3.02 requires a different notion of conceivability, or at least strong and well supported intuitions to the effect that Wittgenstein's notion is inadequate, as many things are intuitively thinkable that are clearly impossible. However, as we shall see, defining an alternative notion of conceivability has proved pretty hard, while attempts at showing, of some situation S, that it is both intuitively conceivable and impossible tend to depend on assumptions that are as theoretically burdensome as the claim that there are necessary facts.

Besides, there appear to be strong general reasons for not wishing to dispense with 3.02 (more generally, with the inference from conceivability to possibility for *some* notion of conceivability); for if we give up the inference, we seem to lose a considerable part of our modal knowledge. Claims that S is possible are often assessed by testing S's conceivability, on the assumption that whatever is conceivable is possible (Yablo, 1993, p. 3). If the assumption is false, conceivability is no longer a safe guide to possibility. What would then be the alternative? Critics of the conceivability → possibility inference have sometimes proposed that knowledge of possibilities could be based on some form of direct modal insight, a rather elusive epistemic notion, unless it is just imagination in disguise.[10]

The second obvious way to get rid of Wittgenstein's argument consists in giving up (6), that is, in holding that P may be a genuine, meaningful sentence even though what it says is inconceivable. In *Naming and Necessity*, Saul Kripke appears to have been of this mind: while he thought it inconceivable that (e.g.) Hesperus is not Phosphorus or that water is not H_2O, he didn't doubt that sentences such as 'Hesperus is not Phosphorus' or 'Water is not H_2O' were perfectly meaningful. If they were meaningless, we could not determine that they are necessarily false.

Prima facie, the view that something strictly and literally inconceivable could nevertheless be meaningfully stated in language looks puzzling. For – it is natural to wonder – if we *understand* a sentence aren't we *conceiving* of its content? Clearly, if the sentence's content is inconceivable it must be that we don't really understand it: the sentence is not really meaningful. However, a moment's reflection shows that we are indeed quite used to regard many sentences as stating inconceivabilities. Take contradictions, both formal ('Socrates is both standing and not standing') and semantic ('Spot A is both red and green all over'); we – or most of us most of the time – tend to think that what they say is inconceivable, yet we also think that they do *say* something. As is well known, Wittgenstein disagreed: he thought that contradictions, like tautologies, were meaningless and said nothing (4.461). Even he, however, felt that a distinction was needed between the kind of meaninglessness that characterizes tautologies and contradictions and the meaninglessness of mere word jumbles: the former are *sinnlos*, the latter *unsinnig* (4.4611). Tautologies and contradictions, as opposed to word jumbles, are 'part of the symbolism' (*ibid.*), that is, they are formed according to the same syntactic rules that generate genuine sentences.

We shall not take up the issue of whether contradictions (or tautologies) are meaningful. That they are not is a direct consequence of the *Tractatus'* conception of sentential meaning (*Sinn*); other theories may disagree. Suffice it to say that many philosophers have found reasons to reject Wittgenstein's views on the issue, so they would not be reluctant to give up (6) and hold that language can indeed state inconceivabilities (e.g., inconsistencies).

Thus, Wittgenstein's argument against necessary facts, as it stands, can be defused simply by admitting that a situation can be both inconceivable and describable in language – not a trifling admission perhaps, but neither a particularly heavy burden, unless one is committed to substantial chunks of Tractarian picture theory. This is not the end of the matter, however, for there is a simpler argument against necessary facts that does not rely on anything like (6).

5 A simpler argument against necessary facts

Suppose we start, like Wittgenstein, with the hypothesis that P describes a necessary fact, so that ~P states something impossible. Assume –again, like Wittgenstein – that nothing impossible can be conceived: it follows that it is inconceivable that ~P. But now, suppose you can show that it is really *conceivable* that ~P (for some P allegedly describing a necessary fact, such as 'Salt is NaCl'): by *reductio*, either P does not describe anything or what it describes is not a necessary fact. This argument is really a class of arguments of the same form. Each argument does not aim at proving that there are no necessary facts, but it shows that some particular sentence P, allegedly describing a particular necessary fact, fails to do so. Thus, arguments of this form are not as general as Wittgenstein's argument. However, if they can be made to work for the canonical examples of necessary facts they are no less devastating than the Tractarian argument.

Versions of this argument form have been widely discussed, in recent years, in the literature concerning conceivability, possibility, and the necessary *a posteriori*. An instance of a particularly simple version, which, however, does not appear to sacrifice anything crucial to the discussion, is the following:

1. For every proposition p, if it is conceivable that p then it is possible that p.
2. It is conceivable that salt ≠ NaCl.
3. So it is possible that salt ≠ NaCl.
4. Therefore it is not necessary that salt = NaCl.

I will refer to arguments of this form as 'the Simple Argument against *a posteriori* necessity'. Notice, first of all, that the Simple Argument is not committed to Wittgenstein's (6), or to the assumption that inconceivable situations cannot be linguistically described. The argument makes no mention of linguistic descriptions: it is about states of affairs (propositions), not sentences. Second, the argument does not explicitly assume that <salt = NaCl>[11] is *a posteriori*. By showing that it is not necessary it shows that, *a fortiori*, it is not both necessary and *a posteriori*. Of course, it could be argued that if <salt = NaCl> were known *a priori* it would not be conceivable that salt ≠ NaCl, so that (2) would be false: this seems to follow from all even minimally demanding notions of conceivability. If so, then (2) presupposes that <salt = NaCl> is known *a posteriori* if it is known at all. However, the argument as it stands is silent about the epistemic status of <salt = NaCl>. The argument's point is that no proposition is necessary if its contradictory is conceivable; it is part of a general strategy that aims at showing that every canonical example of *a posteriori* necessity – every allegedly necessary fact – is such that its contradictory is conceivable. Intuitively, this is just because all such examples are, indeed, *a posteriori* truths: if a situation can only be *discovered* to be the case, then – it is implicitly argued – nothing in what we knew beforehand rules out the possibility that the opposite might have been the case; or, in other words, the opposite was conceivably the case. This is a dubious argument, as we shall see. Anyway, nothing of this is formally assumed in the Simple Argument.

Prima facie, the Simple Argument seems to be open to two challenges, that is,

(a) Reject premise (1): argue that conceivability does not entail possibility;
(b) Reject premise (2): argue that it is not really conceivable that salt ≠ NaCl.

But in fact there is a third possibility:

(c) Reject the Simple Argument as equivocating between different senses of 'conceivable' and 'possible'.

(a) was Hilary Putnam's off-the-cuff choice in 1975 (p. 233), when confronted with the apparently obvious conceivability of (Earthian) water not being H_2O:

> We can perfectly well imagine having experiences that would convince us (and that would make it rational to believe that) water *isn't* H_2O. In that sense, it is conceivable that water isn't H_2O. It is conceivable but it isn't logically possible! Conceivability is no proof of logical possibility.

He argued for his view in (1990). Other philosophers who want to reject the conceivability → possibility inference include Tidman (1994) and Fiocco (2007). Soames also comes out in favor of this strategy, and believes that Kripke ought to (2003, II, pp. 452–3).[12] We already mentioned one theoretical cost of this solution: it appears to deprive us of epistemic access to possibilities, unless we are prepared to countenance the mysterious faculty of modal intuition. We also saw that there are notions of conceivability, such as Wittgenstein's, on which the inference does hold; therefore, this solution is badly in need of an alternative notion of the conceivable (or else, of strong intuitions that some intuitively thinkable situations are nevertheless impossible).

In *Naming and Necessity*, Kripke argued for (b): he claimed that it is not *really* conceivable that, say, water is not H_2O, that is, that it does not have the essence it in fact has. What *is* conceivable is a different state of affairs, that we tend to mistake for the state of affairs that water ≠ H_2O. Namely, we can conceive of our having exactly the evidence we have concerning water – of being 'qualitatively in the same epistemic situation' – except that such evidence is generated by a substance that is not H_2O (and therefore is not water).[13] In the Hesperus/Phosphorus case, we can conceive of having the evidence we do have (which as a matter of fact concerns a single planet, that is, Venus) although *two* planets are involved, at least one of which is not Venus. This, however, would not be conceiving *that Hesperus is not Phosphorus* (Kripke, 1980, pp.103–4). Similarly, conceiving of a waterlike substance that is not H_2O would not be conceiving that *water* is not H_2O.

There are two problems with Kripke's strategy. First, its application may be wider that desirable. Second, an objection due to Scott Soames seems to show that Kripke's strategy involves a misdescription of the function of the evidence on which we determine the truth of (say) 'Water = H_2O'; once the misdescription is corrected, the entire strategy appears to be dispensable.

Strategy (c) was proposed by David Chalmers (2002b) and (to some extent) by Stephen Yablo (2002). It consists of two moves: first, distinguish between different senses of both 'conceivable' and 'possible'; second, show that there is no *uniform* reading of the Simple Argument on which the argument is both valid and sound, that is, no reading on which (i) 'possible' and 'conceivable' are taken in the same sense throughout the argument, (ii) the premises are true, and (iii) the conclusion follows validly.

We shall now examine the three strategies, beginning with (b).

6 Kripke's way out

There is a sense in which supporters of strategy (b), such as Kripke, *prove* that we cannot (and do not) really conceive that – say – salt is not NaCl. Assume that salt is essentially NaCl, so that it is impossible that it is not NaCl; assume further that conceivability entails possibility, so that, by contraposition, nothing impossible can be conceived; it follows that it is inconceivable that salt is not NaCl. Now, essentialism concerning natural kinds is an independently motivated thesis of Kripke's. The conceivability → possibility inference he is not willing to give up, because of the role it plays in the arguments for rigid designation: after all, we tend to believe that Aristotle might not have written the *Metaphysics* or taught Alexander because we can easily *conceive* that he did not.[14] Thus, for Kripke, both premises of the inconceivability claim are well established. However, to avoid his *modus ponens* turning into a *modus tollens* he owes us an explanation of why it *seems to us* that it is conceivable that salt is not NaCl. This is what the argument about epistemically indistinguishable states of affairs is intended to provide. It seems to us that we can conceive of salt not being NaCl because we can conceive of a different, though related state of affairs, namely of a substance – not salt, but epistemically indistinguishable from salt: a 'symptomatic counterpart' of salt[15] – not being NaCl.

Is this explanation satisfactory? Crispin Wright has remarked that it risks depriving the conceivability → possibility inference of all theoretical efficacy: for whenever it is claimed that some state of affairs *p* is conceivable and therefore possible, it can be retorted that *p* merely *looks* conceivable, but it is not (what is conceivable is a different, though epistemically related state of affairs *p'*); so that the conceivability → possibility inference, though valid, can never be safely applied to establish that some state of affairs is metaphysically possible (Wright, 2002, p. 410). Wright further noticed that, however, Kripke himself showed that the strategy is in fact not universally applicable: for there are cases in which we cannot distinguish between conceiving that, say, *a* is not *P* and conceiving of a symptomatic imitation of *a* not being *P*. Such is, for example, the case of pain: as there is no symptomatic imitation of pain (whatever feels like pain *is* pain), we cannot distinguish between genuinely conceiving that *pain* is not the stimulation of C-fibers and conceiving that *some epistemic counterpart of pain* is not the stimulation of C-fibers.[16] Thus, the conceivability → possibility inference is allowed to retain some theoretical efficacy for there are cases in which it can be safely applied – cases in which what looks conceivable *is* indeed conceivable.[17]

Still, even if the strategy is not universally applicable it may be applicable in just too many cases. For example, it could be surmised that when we believe we are imagining George W. Bush not being re-elected President of the United States we are *really* imagining some epistemic counterpart of Bush – someone we would be unable to distinguish from Bush – not being re-elected. Surely here we could not object by claiming that whoever looks (feels, etc.) like G.W.Bush *is* G.W.Bush. So, perhaps Kripke's strategy does weaken the applicability of the conceivability →possibility inference to an undesirable extent.

Another objection to strategy (b) was raised by Scott Soames. To introduce it, let us start with a rather naive remark. Kripke says that the evidence we possess relative to what we call 'water' is compatible with the liquid stuff we are dealing with *not* being H_2O. But then, what justifies our belief that the stuff *is* H_2O? Kripke's claim seems to entail that whatever evidence we have does not discriminate between its being H_2O and its being something else, say XYZ.

Here, the answer is simply that saying that we have discovered that water is H_2O is just saying that we found *additional* evidence that is sufficient to decide the issue in favor of the stuff being H_2O rather than anything else. We can conceive that the stuff is not H_2O because all *previous* evidence is consistent with its not being H_2O; obviously, the evidence we now have, justifying our belief that water is H_2O, is *not* consistent with its not being H_2O. That the stuff is not H_2O was, but no longer is consistent with our epistemic state.

Well and good. But now comes Soames's objection (2003, p. 446, 448). The discriminating evidence is supposed to rule out the possibility that the liquid stuff we are dealing with be other than H_2O. However, if it is H_2O, it is necessarily H_2O; so there is no such possibility to rule out. What is, then, the function of the discriminating evidence? According to Soames, the answer must be that its function is to rule out the *merely epistemic* possibility that water ≠ H_2O. But, Soames adds, once we have the distinction between epistemic and metaphysical possibility we no longer need Kripke's convoluted argument against the conceivability of <water ≠ H_2O> (i.e., against premise (2) in the Simple Argument); a more direct argument is available, which really amounts to rejecting (1) rather than (2), that is, to adopting strategy (a) instead of (b). For Soames is interpreting 'conceivable' as 'epistemically possible', then proceeding to accept (2) (it is conceivable [=epistemically possible] that water ≠ H_2O) while rejecting (1) (the conceivability → possibility inference).

But more on this later. The lesson we can draw at this point is that strategy (b) – Kripke's strategy – is doubly flawed. First of all, it is not clear that it could not be used to undermine some of Kripke's own arguments for the rigidity of proper names, by casting doubts on our ability to really imagine (e.g.,) that *Aristotle* never went into pedagogy or that *George W. Bush* was not re-elected in 2004. Second, it requires the distinction between epistemic and metaphysical possibility if it is to make sense of our coming to know <water = H_2O>, as opposed to other states of affairs that are said to be conceivable on the basis of all previous evidence. But once we have the distinction, Kripke's strategy becomes redundant: a more direct strategy is available.

7 Interpretations of 'conceivable'

Both remaining strategies, (a) and (c), turn on the meaning of 'conceivable'. It will then be helpful to pause and scrutinize some attempts that have been made at clarifying this puzzling notion. Most of them have issued in one of three alternative suggestions: (A) conceivable = imaginable;

(B) conceivable = describable in language; (C) conceivable = epistemically possible.[18]

(A) When 'imaginable' is understood in terms of the ability to form mental images, this interpretation has long been criticized as being too restrictive: many things are ordinarily taken to be conceivable that we cannot form mental images of. For example, we can conceive of, but not form a mental image of a sound that no human ear can hear; or of a kiliagon, that is, a polygon with a thousand sides (as distinct from a polygon with 999 sides); or of God's existence; or of the fact that gold has atomic number 79. In all these cases, criticism of (A) assumes – plausibly – that the things and facts in question are intuitively conceivable, that is, that most of us would have no hesitation to apply the word 'conceivable' in such cases. The intuition may be strengthened by the implicit belief that such things and facts are (at least) *possible*, and what is possible must be conceivable. The latter inference – from possible to conceivable – is usually regarded as much less problematic than the converse, from conceivable to possible; though even there, counterexamples have been proposed.[19] Be that as it may, intuitions are virtually unanimous that we can conceive of much more than we can form mental pictures of.

Recently, David Chalmers has put forth the notion of 'positive conceivability': a kind of imaginability he wants to keep distinct from image-ability or 'perceptual imaginability', that is, the notion we have just been discussing. Neither does positively conceiving that *p* coincide with entertaining the proposition that *p*, for there are propositions (e.g., 'complex unknown mathematical propositions') that are not easy to imagine (2002b, p. 151). Positively conceiving that *p* seems to be, imagining things and events that one might think of describing by '*p*': for example, positively conceiving of Germany winning the last World War would be imagining 'a world in which certain German armies win certain battles and go on to overwhelm Allied forces within Europe' (*ibid.*). But of course, that such events took place is neither necessary nor sufficient for Germany to win the last World War (the Germans might have lost those particular battles and yet the war be decided in favor of Germany by the Japanese armies or by some catastrophic natural event, or they might have won those battles and overwhelmed the Allies in Europe but eventually lost the war because of allied victories everywhere else and economic collapse in Europe). Thus one may doubt that imagining such events really amounts to conceiving of Germany's victory in the last World War. Moreover, what does imagining the German armies winning certain battles amount to, if it is neither mentally picturing the event nor entertaining the proposition that it takes place? There is strong reluctance to accept the idea that imagining that *p* may just amount to thinking of the sense of *p*, which, in turn, may or may not involve the formation of mental images. Chalmers's proposed notion of positive conceivability (which, as he admits, 'resists straightforward definition' (*ibid.*)) is an expression of such reluctance.

(B) If conceivable = describable in language, then, as we noticed above, conceivability does not entail possibility, as contradictions of all sorts are linguistically describable. To salvage the conceivability → possibility inference one would have to claim, with Wittgenstein's *Tractatus*, that contradictions are meaningless: no inconsistent situation is meaningfully describable in language. We noticed that many philosophers are inclined to resist this thesis. However, would they also reject it *when describability in language is taken as an* explicatum *of conceivability?* In other words, would they be prepared to claim *both* that 'conceivable' means 'describable in language' *and* that any inconsistent situation is perfectly describable in language? Take Chomsky's 'Colorless green ideas sleep furiously'. Could the situation it describes be regarded as (intuitively) conceivable? If not, there are three options: either (a) conceivability does not amount to linguistic describability, or (b) Chomsky's 'sentence' is not really a sentence of English, or (c) not every linguistic description will do: to say that S is conceivable is to say that it can be *coherently* described in language. Hence, partisans of definition (B) must either specify restrictions on what counts as a genuine linguistic description (beyond syntactic and lexical specifications) or qualify 'describable' as '*coherently* describable'. It is not clear whether, once such restrictions and qualifications have been added, conceivability would entail possibility. For some choices, it clearly would: for example, if a description is taken to be coherent if and only if it has a model, then whatever is coherently describable is possible.

(C) *p* is said to be epistemically possible if and only if it is not ruled out by what we know, that is, iff it is possible 'for all we know'. If conceivability is identified with epistemic possibility, it seems that the relevant knowledge must be restricted to *a priori* knowledge, for otherwise it would be inconceivable that the Germans won the last World War or that the French flag were white, orange and green.[20] Aside from being deviant with respect to common usage, such consequences would, again, endanger the possibility → conceivability inference, as all such states of affairs appear to be unproblematically possible. Skipping the restriction, on the other hand, would yield a cheap victory against the Simple Argument, as it is not compatible with what we now know, therefore not conceivable, that salt ≠ NaCl (thus (2) would fail). A high price to pay to get rid of the Argument.

However, if the epistemically possible is expanded to whatever is consistent with what we know *a priori*, the notion (and conceivability with it) inherits the relative indeterminacy of the *a priori*. We cannot confine *a priori* knowledge to logic and mathematics, for we don't want it to be conceivable that bachelors are married or that red is green; but as soon as we make semantic competence constitutive of *a priori* knowledge we have to face the notorious problem of delimiting semantic competence with respect to world knowledge. The notion of conceivability will then be as fuzzy as the boundary between them. Nevertheless, a fuzzy notion is better than nothing: consequently, (C) has proved popular as a definition of conceivability.

8 Is the Simple Argument based on equivocation?

David Chalmers (2002b) has argued that there are senses of 'conceivable' and 'possible' such that both (1) and (2) in the Simple Argument are true, but, taking 'possible' in that sense, either (3) does not entail (4) or (4) is not the negation of the intended claim, that there are (metaphysically) necessary *a posteriori* truths and that salt = NaCl is one of them. In the former case the argument fails, while in the latter it is an *ignoratio elenchi*: it doesn't prove what it intended to prove. Yablo (2002) can be read as working along similar lines: he wants to define a notion of possibility such that (1) is plausible, but – understanding 'possible' in that sense – (4) does not negate the Kripkean claim.

As is well known from previous writings (Chalmers, 1996, 2002a), for Chalmers there are two notions of possibility (and, correlatively, of necessity). *Secondary* possibility is the familiar notion of metaphysical possibility: *p* is secondarily possible iff there is a counterfactual world such that *p*. Given that salt is NaCl, it is (secondarily) impossible that salt ≠ NaCl. The other notion, *primary* possibility, is based on the idea of the actual world being different from what we – *a posteriori* – know it to be. For all we know *a priori*, our world, that is, the actual world, could be such that salt is not NaCl. 'Salt' might – for all we know *a priori* – designate a substance that is not NaCl. In this sense, it is possible (*primarily* possible) that salt ≠ NaCl. So primary possibility is just epistemic possibility (expanded to consistency with a priori knowledge). (Chalmers, 2002b, p.157).

The matter of conceivability is more intricate. Here Chalmers introduces three distinctions. First, between negative conceivability

(= compatibility with what we know *a priori*) and positive conceivability (= imaginability, though not just image-ability, as we saw above). Secondly, between *prima facie* conceivability and ideal conceivability, that is, conceivability upon ideal rational reflection (a mathematical proposition that was discovered, after centuries of research, to be an impossibility may have been *prima facie* conceivable, but it is not, and never was ideally conceivable). The third, crucial distinction is between *primary* and *secondary* conceivability: *p* is primarily conceivable iff it is (negatively or positively, *prima facie* or ideally) conceivable that *p is actually* the case. It is secondarily conceivable iff it is (negatively or positively, etc.) conceivable that *it might have been* the case (2002b, p. 157).

Chalmers's distinctions combinatorially originate a large number of pairings between notions of conceivability and notions of possibility. We shall not explore all alternatives, nor does Chalmers. From the standpoint of the present discussion, his point is that, though there are notions of both conceivability and possibility on which the inference from conceivability to possibility goes through – that is, premise (1) in the Simple Argument holds – *and* it is conceivable that salt ≠ NaCl (premise (2) holds), none of the relevant notions of possibility is such that the conclusion follows that it is not necessary that salt = NaCl, *when 'necessary' is interpreted as standing for metaphysical, or 'secondary' necessity* (as it supposedly should be). On the other hand, there are *other* notions on which the conceivability → possibility inference also holds, and the conclusion that it is not necessary that salt = NaCl *would* follow, if it were conceivable that salt ≠ NaCl; but, on the relevant notion of conceivability, it is not. If Chalmers's analysis were convincing, he would have succeeded in defeating the Simple Argument without sacrificing the conceivability → possibility inference.

However, his analysis is not entirely convincing, and not just because he does not seriously scrutinize every combinatorial alternative generated by his own distinctions. In fact, most such distinctions are really idle wheels in Chalmers's strategy against the Simple Argument (which is not to deny that some of them may be interesting and useful in other respects). Actually, all the work is done by the following three claims:

(c.1) Primary conceivability entails primary possibility,
(c.2) Secondary conceivability entails secondary possibility,
(c.3) Primary conceivability does not entail secondary possibility.

According to Chalmers, it is primarily but not secondarily conceivable that salt ≠ NaCl. As <salt ≠ NaCl> is primarily conceivable, it is primarily possible (by c.1); it is not, however, secondarily possible as well (this is ruled out by c.3). As it is not secondarily possible that salt ≠ NaCl, it does not follow that <salt = NaCl> is not (secondarily, i.e., metaphysically) necessary. Thus the Simple Argument's conclusion is not reached. On the other hand, if it were secondarily conceivable that salt ≠ NaCl, then it would be secondarily possible (by c.2) and the Argument's conclusion would follow; but, as we premised, it is not secondarily conceivable that salt ≠ NaCl. This is how Chalmers shows that the Simple Argument is fallacious by equivocation: the Argument only concludes by equivocating on 'conceivable' in premises (1) and (2). The Argument (repeated below) takes 'conceivable' in (2) to be 'primarily conceivable', so that (2) is true; but then, in order to reach conclusion (4), it must take 'possible' in (1) and (3) to be 'secondarily possible'. However, if 'possible' in (1) is 'secondarily possible' then (1) is only true if 'conceivable' is 'secondarily conceivable', not 'primarily conceivable' as it had been taken to be to make (2) true.

1. For every proposition p, if it is conceivable that p then it is possible that p.
2. It is conceivable that salt ≠ NaCl.
3. So it is possible that salt ≠ NaCl.
4. Therefore it is not necessary that salt = NaCl.

Chalmers' argument depends on his three claims, (c.1)–(c.3), and two premises: that allegedly necessary *a posteriori* truths are primarily conceivable, and that they are secondarily inconceivable. Chalmers's reasons for both premises are essentially reformulations of the Kripkean intuitions. It is not secondarily conceivable that, say, Hesperus is not Phosphorus because, assuming rigid designation, no counterfactual situation we can imagine (in any sense) is a situation in which Hesperus is not Phosphorus (2002b, pp. 158–9); by contrast, it is primarily (positively) conceivable that Hesperus is not Phosphorus because we can imagine situations that would be described as situations in which Hesperus is not Phosphorus (e.g., that there are two distinct celestial bodies, respectively named 'Hesperus' and 'Phosphorus', generating the experiences we do have, etc.), and nothing in what we know *a priori* rules out that some such situation may actually obtain. Discounting scepticism about the notion of positive conceivability, it seems that Chalmers's premises are just as solid as the analysis on which Kripke's version of strategy (b) is based.

Claims (c.1)–(c.3), on the other hand, are somewhat opaque, as they depend on notions – primary and secondary conceivability – that are themselves defined in terms of other notions of conceivability. For example, as we saw above, *p* is primarily conceivable iff it is (negatively or positively, *prima facie* or ideally) conceivable that *p* is actually the case. Thus, in a way, there is no such thing as unqualified primary conceivability: there is primary *positive* conceivability, primary *negative* conceivability, etc. Claims (c.1)–(c.3) ought then to be evaluated for each such interpretation of conceivability: for example, we may ask whether from the fact that we can positively conceive (i.e., 'imagine') that the actual case is that water is not H_2O it follows that the actual case might indeed be that water is not H_2O. I, for one, wouldn't know how to answer such a question, as I don't know what it is to imagine that water is not H_2O, as distinct from entertaining the hypothesis that water is not H_2O. Chalmers himself goes to great lengths in exploring some of the connections between notions of conceivability and notions of possibility, but I find his discussion inconclusive.

Therefore, in spite of its clever rhetoric, strategy (c) is actually in a worse shape than strategy (b) as an objection to the Simple Argument. For it shares with strategy (b) reliance on the Kripkean analysis of conceivability with all its problems, while it needs, in addition, to establish inferential connections, positive or negative, between *some* notion of conceivability, on the one hand, and possibility (primary and secondary) on the other, that have not been satisfactorily established so far.

9 Rejecting the inference from conceivability to possibility

Rejecting the conceivability → possibility inference – strategy (a) – is a costly move, as it deprives us of our traditional access to modal knowledge.[21] Nevertheless, some philosophers have chosen this option.

Paul Tidman (1994) put forth an argument against the inference based on an interpretation of conceivability that is akin to, though not identical with interpretation (C): conceivable = believable. Grant that it is *prima facie* plausible that we can believe (i.e., conceive) impossibilities: for example, that this particular cat is a robot.[22] As this *cat* couldn't be a robot, the inference to possibility does not go through, at least *prima facie*. To maintain that in fact it does, we have to show that, appearances to the contrary, we cannot believe impossibilities (i.e., we must show that the inference's premise is not satisfied: we cannot really believe that this cat is a robot). But how could we? We are assuming that (at a minimum) it seems to us that we can believe that this cat is a robot. If we could tell that we cannot, it wouldn't seem to us that we can. So, either we can believe impossibilities, or, if we cannot, we cannot show that we cannot. Hence, either the inference is invalid or we cannot show that it is valid by ruling out apparent counterexamples.

It could be remarked that, even if we are in no position to *show* that we are not really conceiving that *p*, it can nevertheless be the case that *we aren't* really conceiving that *p* (we might be under the inevitable delusion that we can). But aside from that, Tidman's argument appears to assume the illegitimacy of Chalmers's distinction between *prima facie* conceivability and ideal conceivability. Why couldn't I, upon reflection, come to the conclusion that what I was conceiving (i.e., what I thought I could believe) is not best described by 'this cat is a robot' but, say, by 'this cat-like object is a robot'? Tidman appears to be assuming that I couldn't have reasons to choose the latter description over the former. This, however, amounts to assuming that we cannot rationally reflect upon the content of our conceivings, that is, that we can't go beyond *prima facie* conceiving; which is both arbitrary and implausible.

Before Tidman, Hilary Putnam (1990) argued for the invalidity of the inference (he had taken such invalidity for granted in his (1975)). Putnam relied on Kripke's notion of an epistemic counterpart, but he put it to a very different use. In Kripke's view, when we say that it is conceivable that salt is not NaCl we are *mis*stating something else, namely that it is conceivable that there is an epistemic counterpart of salt which is not NaCl. This is indeed conceivable, according to Kripke; but then, what we are asserting to be conceivable is *not* that salt ≠ NaCl. Putnam thinks that 'it is conceivable that salt is not NaCl' *means* that some epistemic equivalent of salt may not be NaCl: 'What is conceivable is a matter of epistemology' (1990, p. 62). Kripke used the notion of an epistemic counterpart to explain how *it may* (wrongly) *seem to us* that it is conceivable that salt ≠ NaCl; Putnam uses it to *explicate* 'it is conceivable that salt ≠ NaCl'.

Having assumed that this is what 'conceivable' means, it seems to follow that conceivability does not entail possibility. It stands to reason that, *if* what we mean by saying that it is conceivable that salt is not NaCl is that *something*, not necessarily salt but something epistemically indistinguishable from salt, might not be NaCl, then this does not entail that *salt itself* might not be NaCl. So conceivability does not entail possibility. The issue is, of course, whether we got the meaning of 'conceivable' right. Is conceivability 'a matter of epistemology', as Putnam says?

Clearly, Putnam's understanding of conceivability amounts to assuming a version of definition (C) ('conceivable' = 'epistemically possible').

For suppose it is epistemically possible that salt ≠ NaCl, that is, suppose that our (qualitative) epistemic situation is consistent with <salt ≠ NaCl>. This is to say that what we know about salt does not rule out the hypothesis that it is not NaCl, that is, that some epistemic equivalent of salt might not be NaCl, which is what Putnam means by 'It's conceivable that salt ≠ NaCl'. And conversely, if some epistemic equivalent of salt might not be NaCl then nothing we know about salt rules out the hypothesis that it is not NaCl, etc. (As we noticed above (see p.161), the relevant knowledge must exclude *a posteriori* knowledge, for once we *know* (*a posteriori*) that salt *is* NaCl, then something we know about salt *does* rule out the hypothesis that it is not NaCl).[23]

We saw that definition (C) of 'conceivable' runs into some problems due to the fuzzy boundaries of *a priori* knowledge. When definition (C) is assumed in attempting to refute the conceivability → possibility inference, however, an additional problem arises: if conceivability is understood as per (C), rejection of the inference turns out to be equivalent to the assumption that there are necessary *a posteriori* truths.

The proof is easy. First, suppose that conceivability$_{(C)}$ does not entail possibility, so that, for some proposition *p*, *p* is both conceivable$_{(C)}$ and impossible. Thus ~*p* is necessary. Now suppose it were *a priori* that ~*p* is necessary: then it would not be conceivable$_{(C)}$ that *p*, for if *p* were consistent with what we know *a priori* , it would not be *a priori* that ~*p* is necessary. But we assumed that *p* is conceivable$_{(C)}$. Hence, that ~*p* is necessary can only be known *a posteriori* (if at all). Thus, if conceivability$_{(C)}$ does not entail possibility there are necessary truths that can only be known *a posteriori*.

For the other direction, suppose there are necessary truths that can only be known *a posteriori*, and let *p* be one such. As *p* can only be known *a posteriori*, ~*p* is consistent with what we know *a priori*. Hence ~*p* is conceivable$_{(C)}$. But *p* is necessary, therefore ~*p* is impossible: so, conceivability$_{(C)}$ does not entail possibility.[24]

That such an equivalence holds is, upon reflection, fairly obvious. Suppose we identify conceivability with epistemic possibility. If we tend to think that conceivability, so understood, does not entail possibility that must be because we feel that metaphysical modality is independent of what we know *a priori*: we may not know *a priori* whether P is possible, impossible, or necessary. For example, there may be necessary truths that cannot be known *a priori*. If we didn't have such an intuition, we would have no reason to reject the inference from epistemic possibility (i.e., conceivability) to metaphysical possibility. We have no *independent* intuition militating against the inference: if we reject it, that is because we are already convinced that there may be necessary facts. If conceivability is understood as epistemic possibility, rejecting the inference *derives* from the assumption that modality is independent of *a priori* knowledge.

A man's *modus ponens* is another man's *modus tollens*. The Simple Argument shows that, as (among other things) conceivability entails possibility, there are no necessary facts. The friend of necessary facts retorts that, if we assume (*inter alia*) that there may be necessary facts we will see that the conceivability → possibility inference fails, so that the Simple Argument does not go through. As there seems to be no independent argument against the inference, what we have here is just a clash of intuitions. If conceivability is equated with epistemic possibility, the Simple Argument can only be refuted (within strategy (a)) by assuming from the start that its conclusion is mistaken.

10 Conclusion

The case for necessary facts is not as strong as it is usually taken to be. A simpler and less committal variant of Wittgenstein's original argument – the Simple Argument – is not easily refuted. The most popular strategy against the Argument – Kripke's strategy (b) – turned out to be most problematic. Chalmers's strategy (c) shares its main weaknesses; moreover, it depends on notions of conceivability that are either obscure or unconvincing. Strategy (a) – rejecting the inference from conceivability to possibility – depends on a suitable definition of the notion of conceivability (for on Wittgenstein's notion of conceivability the inference does hold). However, we saw that definition (A), conceivability as image-ability, is either too restrictive or (in Chalmers's version) obscure. Definition (B), conceivability as linguistic describability, will make the inference invalid only if it is admitted that contradictions are describable. This is fine with many philosophers on the face of it; however, one doubts that those same philosophers would still agree *when linguistic describability is taken as a definition of conceivability*. It is not clear that one can hold *both* that contradictions are linguistically describable *and* that conceivability can be defined as linguistic describability.

Definition (C), conceivability as epistemic possibility (with the usual restriction to *a priori* knowledge), does the job of invalidating the conceivability → possibility inference only on the assumption that modality is independent of *a priori* knowledge. However, this entails assuming that there may be necessary facts. But then, what we are doing is not so much refuting the Simple Argument as showing that an alternative conception is coherent: if we *assume* that there are necessary facts we shall not get in trouble with the Simple Argument, for one of its premises – the conceivability → possibility inference – will fail. That is, it will fail *if* we take conceivability to be epistemic possibility *and*, in addition, we rely on the assumption that modality is independent of *a priori* knowledge. Of course, partisans of the Simple Argument needn't go along with either assumption.

Consequently, if the case for direct, nonderivative metaphysics only rested upon the rejection of the Simple Argument against necessary facts it would be a rather weak case. It appears that (as far as the present dialectic is concerned) although necessary facts can be coherently postulated, contrary intuitions concerning both conceivability *per se* and its relation to possibility are not easily dispelled. The most promising attempts at undermining the latter rest upon the assumption that there are necessary facts, while Kripke's strategy against the conceivability of <water ≠ H_2> (and the like) is shot through with serious difficulties.

There are, of course, other arguments for direct metaphysics. Even the existence of necessary facts can be defended, for instance, by appealing to rigid designation and the ensuing necessity of <Hesperus = Phosphorus>, or to essentialism and the ensuing necessity of <water = H_2O>. However, arguments for rigid designation and essentialism *per se* will not yield a notion of conceivability, nor will they determine failure of the conceivability → possibility inference. Thus, the Simple Argument stands.

This is not to say that the Wittgensteinian legacy is vindicated. As we saw, Wittgenstein's original argument against necessary facts depends on assumptions, such as the picture theory of language, that few would be prepared to go along with. However, Wittgenstein's resistance against necessary facts and many of his argumentative moves still deserve to be taken seriously.[25]

Notes

1. This line of thought – the 'semantic tradition' – was presented and discussed by Alberto Coffa (1991).
2. I will sometimes use 'conceptual scheme' as shorthand for 'conditions of experience', or 'language', or 'framework', or 'form of representation', depending on which philosopher I am talking about.
3. Kant 1787, Kemp Smith translation, p. 239.
4. Kemp Smith translation, p. 239.
5. 'Durch eine optische Täuschung scheinen wir im Innern der Dinge zu sehen, – was auf unsrer Brille gezeichnet ist. 'Es ist doch so –' sagen wir uns

wieder und wieder. – Es ist uns, als müssten wir das Wesen der Sache erfassen, wenn wir unsern Blick nur ganz scharf auf dies Faktum einstellen, es in den Brennpunkt rücken könnten [= PI §113]. Denn es scheint eben im Innern der Sache zu liegen. Erst wenn diese optische Täuschung entfernt ist können wir nun die Sprache einfach sehen, wie sie ist.// Der Ausdruck dieser Täuschung aber ist die metaphysische Verwendung unser Wörter. Denn man prädiziert nun von der Sache, was in der Darstellungsweise liegt. Die Möglichkeit des Vergleichs, die uns beeindruckt nehmen wir für die Wahrnehmung einer höchst allgemeinen Sachlage.

6. Roughly corresponding to the first half of the pre-war *Investigations*.
7. *Critique of Pure Reason*, First Analogy, Kemp Smith tr. p. 212.
8. For many philosophers in the tradition (e.g., for Carnap, C.I. Lewis, and the later Wittgenstein), though not for Kant himself.
9. This notion of thought is not easily reconciled with Wittgenstein's view that individual thoughts are configurations of 'psychic elements that have with reality the same kind of relations as do words' (*To Russell*, August 19, 1919). At least for some pictures – paintings, say – it is not clear how they could be configurations of psychic elements. For an able though not entirely convincing attempt at reconciliation see Frascolla, 2007, pp. 37–40.
10. See for example, Tidman, 1994 and Fiocco, 2007; they both refer back to Van Cleve, 1983.
11. I adopt the convention on which <p> = the proposition that p.
12. Yablo (1993) grants that there are exceptions to the inference's validity, while still taking conceivability to provide *prima facie* evidence of metaphysical possibility (p. 34).
13. Kripke, 1980, pp. 142–3. His discussion concerns gold not being an element but a compound.
14. See Yablo, 1993, p. 3.
15. Wright, 2002, p. 414.
16. See Kripke, 1980, p. 151; Wright, 2002, pp. 410–1.
17. In (1992), Wright introduced a different strategy for doubting the genuine conceivability of a state of affairs, one that – contrary to Kripke's – would be universally applicable. For an objection to the latter claim see Marconi, 2008, pp. 188–9.
18. See the discussion in Gendler Szabó and Hawthorne, 2002, pp. 7–26.
19. Yablo claimed that neither Goldbach's conjecture nor its negation is intuitively conceivable, though one or the other must be true and therefore possible (1993, p. 11). However, he may have been using 'conceivable' in the sense of 'imaginable', in which case his example would be just one more counterexample to (A), rather than against the inference from possible to conceivable.
20. Gendler Szabó and Hawthorne (2002b) introduce a distinction between two notions of epistemic possibility: on the *permissive* notion, *p* is epistemically possible for X iff X does not know that ~*p*; on the *strict* notion, it is epistemically possible iff it is consistent with all that X knows. *p* may be permissively possible without being strictly possible (for some X): for example, Goldbach's conjecture might be inconsistent with what I know (hence not strictly possible) though I don't know it to be false (so it is permissively possible). I what follows, I will take the notion in the strict sense. Gendler

Szabó and Hawthorne (p. 4) also note that in neither sense does conceivability entail epistemic possibility, for the reason highlighted in the text.

21. See Yablo, 1993 and Gendler Szabó and Hawthorne, 2002 on Hume's and Descartes' use of the inference and contemporary objections to such use.
22. Not Tidman's example in this context.
23. Actually, regarding individual propositions such as <salt ≠ NaCl> a posteriori knowledge could be included, provided it does not entail that salt = NaCl.
24. We can likewise prove that the thesis that conceivability$_{(C)}$ entails possibility is equivalent to the claim that there are no necessary *a posteriori* truths. Suppose conceivability$_{(C)}$ entails possibility, and suppose (for *reductio*) that *p* is both necessary and *a posteriori*. As *p* is necessary, ~*p* is impossible, so, by the hypothesis, it is inconceivable$_{(C)}$, that is, inconsistent with what we know *a priori*: therefore p can be known *a priori*. Conversely, suppose that every necessary truth is *a priori*, and assume (for *reductio*) that the conceivability$_{(C)}$ → possibility inference does not hold, that is, for some *p*, *p* is both conceivable$_{(C)}$ and impossible. As *p* is impossible, ~*p* is necessary. Then by the hypothesis, ~*p* can be known *a priori*, so *p* can be ruled out *a priori*, that is, *p* is not conceivable$_{(C)}$.
25. Thanks to Guido Bonino, Pasquale Frascolla, and Alberto Voltolini for comments on a previous version.

8

Wittgenstein and Frege on Proper Names and the Context Principle

Eva Picardi

Consider the following three sentences:

1. Nur im Zusammenhange eines Satzes bedeuten die Wörter etwas. Es wird also darauf ankommen, den Sinn eines Satzes zu erklären, in dem ein Zahlwort vorkommt. (Frege, *Grundlagen der Arithmetik* §62)[1]
2. Nur der Satz hat Sinn; nur im Zusammenhange des Satzes hat ein Name Bedeutung. (Wittgenstein, TLP 3.3)
3. Das Benennen ist noch gar kein Zug im Sprachspiel ... Man kann sagen: Mit dem Benennen eines Dings ist noch *nichts* getan. Es *hat* auch keinen Namen, außer im Spiel. Das war es auch, was Frege damit meinte: ein Wort habe nur im Satzzusammenhang Bedeutung. (Wittgenstein, PI §49)

Their English translations read as follows:

1. (CP 1) Since it is only in the context of a proposition that words have a meaning, our problem becomes this: To define the sense of a proposition in which a number word occurs.
2. (CP 2) Only propositions have sense; only in the nexus of a proposition does a name have meaning.
3. (CP 3) Naming is so far not a move in the language-game. ... We may say: *nothing* has so far been done, when a thing has been named. It has not even *got* a name except in the language-game. This was what Frege meant too, when he said that a word had meaning only in the context of a sentence.

Is one entitled on the basis of (1) and (2) to come to the conclusion that Frege and Wittgenstein stood to each other in the relation of samesaying as regards the Context Principle? Can one use (3) as evidence in favour of this claim? A moment reflection shows that it would be rash to jump to such a conclusion. And not only because attending to the mere wording of isolated sentences would be sinning against the Context Principle itself, but also because both the Context Principle and the notion of context admit of many different interpretations.

The structure of my paper is as follows: After touching on the different notions of context involved in (CP 1), (CP 2), and (CP3), I will examine Travis's interpretation of (CP 3) against the background of current discussions on contextualism in the philosophy of language. In the second part of my chapter I will examine the bearing of (CP 1), (CP2) and (CP 3) on specific issues relating to the meaning and use of proper names. The view I will argue for is that the later Wittgenstein, unlike the author of TLP, has many interesting things to say concerning the meaning and use of proper names that are not (not all of them, at any rate) at variance with Frege's account of the sense and reference of proper names. My surmise is that if we bear in mind the paramount importance of context in *communication*, we may come to see Frege's and Wittgenstein's observations on the fluctuating meaning of proper names from a new perspective.

1 Varieties of context

As is well known, there is a lively debate going on nowadays between Radical Contextualists, such as Charles Travis and François Recanati, and Semantic Minimalists, such as Ernest Lepore and Herman Cappelen, concerning the role that context is supposed to play in determining the content of what is said in uttering a token of a sentence type on a given occasion. The received model of truth-theoretical semantics seems to totter under the perceived (or imagined) pervasiveness of context sensitivity. Context sensitivity is not confined to indexicals and demonstratives, but affects in subtle ways the intended interpretation of most of our utterances.

No wonder that Wittgenstein's work is often cited either as an early example of an approach to meaning and understanding that allots to context the crucial importance that it richly deserves, or as an example of the nihilistic consequences in store for those who succumb to the temptation of erasing the semantics/pragmatics divide. In his book of 1989, *The Uses of Sense*, Charles Travis who, according to current classifications counts as a Radical Contextualist, traces back to Wittgenstein many ideas that are taken up again in contemporary debates. If Travis is right, even regarding humble sentences such as 'There is milk in the refrigerator' one can at best ask what is conveyed by uttering it on a certain speaking. There is no 'proposition' in the context of which a word can be said to have a meaning. There are at best situated utterances with a *point*.

In recent debates the word 'context' is employed in such a way as to cover both the information derived from external circumstances of utterance relevant to the interpretation of sentences with indexicals and demonstratives, and the inferential guesswork in which speaker and hearer engage on a given episode of conversation, in order to figure out what is conveyed by an utterance over and above its truth conditional core (the 'minimal proposition' expressed by the sentence). It may be well to bear in mind from the start that Wittgenstein used different words for alluding to what falls under this broader notion of context, such as 'Umgebung,' 'Situation,' 'Kontext.' Frege had no special terminology for singling out the contribution that the circumstances of utterance make to the thought expressed by a sentence with indexicals on a certain occasion of its assertoric use. He tended to use the word 'completion' (*Ergänzung*) to cover both the hints dropped by a speaker when addressing a given audience and the contribution that information gathered from external circumstances makes to the thought expressed by a sentence with indexicals on a given occasion.[2] This broader notion of context was mentioned by Frege in many of his writings, and considered in greater detail in the First Logical Investigation of 1918. Frege was eager to show that a sentence with indexicals does not change its truth value depending on the occasion of utterance. The same-sounding sentences express different thoughts, depending on the way the gaps signalled by indexicals and demonstratives are filled in on different occasions of utterance. Thoughts, not sentences, are the primary bearers of truth.

Neither (CP 1) nor (CP 2) relate to this broader notion of context. What is here under discussion is *sentential context*. (CP 1) calls attention, among other things, to the explanatory priority of a sentence over its component words. The leading insight is that in order to provide a semantic account of the way in which words belonging to different syntactic categories may contribute to the meaning and truth conditions of the large variety of sentences constructible in a language; one cannot proceed piecemeal, in a word by word fashion. In the *Foundations of Arithmetic*, where for the first time the Context Principle is formulated, Frege is concerned with a special kind of statements, that is, statements of number. One of the main claims of that work is that number words are proper names of *logical objects*. If by means of purely logical laws and definitions we succeed in fixing the sense and truth conditions of a special kind of sentences, that is, identity statements, in the context of which number words occur , the prospects of logicism as a foundational programme will be greatly enhanced. Frege's agenda was very familiar to young Wittgenstein. However, TLP while endorsing (CP 1), emphatically denies that in arithmetical equations the symbol of identity should be construed along the lines suggested by Frege (1893).[3] In fact, propositions of mathematics do not express thoughts (TLP 5.21), and in this respect differ dramatically from propositions in which names occur as representatives of objects in states of affairs.

In his masterpiece of 1884 Frege had also voiced the hope that scrupulous observance of (CP 1) may prevent us from falling prey both to ideational and to 'physical' conceptions of meaning. According to the former the meaning of a word is a mental image, according to the latter the meaning is the physical object a word in isolation stands for.[4] In addition, Frege believed that (CP 1) may also enable us to keep clear of the many pitfalls of psychologism. One of the most serious defects of psychologism is that it stands in the way of a correct construal of the notion of *objectivity* appropriate to the contents expressed by sentences accessible to (rational) language users (*Foundations*, §26).[5] In the course of his philosophical development Frege supplemented (or replaced, depending on one's interpretation of Frege's work) the Context Principle with Principles of Compositionality and Functionality.[6] Possibly, it is to the latter principle that Wittgenstein approvingly alludes in (CP 2). If this suggestion is correct (cf. TLP 3.318, 4.02, 4.024, 4.027) the reasons why (CP 1) is quoted with approval in PI §49 must be very different from the reasons that led young Wittgenstein to endorse (CP 1).

PI §49 alludes to a notion of context that differs not only from that of *sentential context* implicit in (CP 1) and (CP 2), but also from the notion of context appealed to by Frege (1918) and deployed in recent debates on the semantics/pragmatics divide. What (CP 3) intimates, among other things, is that the significance of the name giving practices available in a language is not exhausted by an appeal to the act of naming. The act of naming is preparatory to the playing of various language games where the words thus introduced are put to a specific use. Frege's Context Principle is mentioned in §49, *after* the notion of a language game has been introduced. What Wittgenstein wants the reader to appreciate at this stage of PI is how vastly Augustine overestimated the significance of ostensive explanations. Even a simple game such as the one played by Builders presupposes for its intelligibility a background of activities in which the ostensive training is embedded:

> With different training the same ostensive teaching of these words would have effected a quite different understanding.
>
> 'I set the brake up by connecting up rod and lever.' Yes, given the whole of the rest of the mechanism. Only in conjunction with that is it a brake-lever, and separated from its support it is not even a lever; it may be anything or nothing (PI §6)

Wittgenstein's emphasis on *application* echoes Frege's injunction in § 46 of *Foundations of Arithmetic* to consider number in the context of a judgement where its primary method of application is prominent. To be sure, both in TLP and in his later writings on the philosophy of mathematics Wittgenstein criticized the *specific* answer given by Frege to the question he had raised. But there can be no doubt that Frege had asked the right question. Indeed, Frege's insight that meaning and *application* should be investigated together was to become a leading theme of Wittgenstein's philosophy in its entirety. What after 1921 underwent dramatic changes is Wittgenstein's conception of what counts as an 'application' of a word or sentence, and how the meaning of a word relates to its various applications (in the plural). The echo of PI §6 can be heard in PI §49. What matters here is not so much the *linguistic* context (the different *kinds* of words and the different *roles* that the same word can play in the context of various sentences), but the web of *activities* in which words are embedded. The ostensive teaching of the word 'slab' does not take place in a void, but is part of a certain activity (the building of a house) in the context of which the shouting of 'Slab' brings about the desired *understanding*. Conceivably, what holds good for common nouns such as 'Slab,' 'Pillar' etc., also applies to proper names such as 'Moses,' 'Nothung,' 'N.N.'

2 Understanding in context

Travis (2009) offers an interesting interpretation of PI §49. He argues that one of the reasons why Frere's emphasized the priority of a sentence over its component parts is because he was eager to stress the possibility of a sentence's multiple decompositions. The 'elements' thus recovered have no other function but that of contributing to the performance of the subtasks on which the truth of the sentence turns on a given occasion. As applied to naming, what Frege's Context

Principle intimates is that naming cannot be done in isolation, but only in the process of saying something that represents what has been named *as being* in a certain way. But we cannot tell in advance, just on the basis of the sentence's grammatical composition, what sort of representing is appropriate under specific circumstances. Possibly Wittgenstein may have found *this* aspect of Frege's Context Principle worth stressing.

Now, the carving-up metaphor, upon which the idea of the multiple decomposition of a sentence into function and argument rests, does indeed play a crucial role both in Frege's account of predication (cf. Picardi, 2008), and in his account of the logical form that can be read into statements of number (cf. *Foundations of Arithmetic* §§62–8). However, in neither case does occasion sensitivity play a role. Moreover, in PI not even a vestige of Frege's distinction of the different role that proper names and concept words play in the context of a sentence is to be found. Also in TLP Frege's way of drawing the distinction between concept and object hardly plays a role. But this is one of the fundamental issues on which (CP 1) is supposed to shed light. And Frege accordingly tries to show that we come to appreciate the difference between concept and object by attending to the different role that singular expressions and concept words play in the context of a sentence.

More generally, it seems to me that the inferential potential of sentences belonging to Frege's Conceptual Notation is sensitive to context only to a very limited extent. Whether on a certain occasion we *focus* on this or that way of splitting a sentence into function and argument may depend on the deductive task at hand. However, the content of the sentence must be *intrinsically* capable of multiple decomposition. Moreover, it is hardly credible that upon hearing a sentence on a certain occasion, one hits straight off on one of its several possible decompositions (with the attending reconceptualizations) into function and argument. This would rob deductive argument of its informative potential.[7]

Travis's book of 2006, *Thought's Footing*, is devoted to a comparison between Wittgenstein and Frege on issues of truth, reference and understanding. Travis's main thesis it that 'propositions' do not determine, in and of themselves, what makes them true or false. Much depends upon particular circumstances of utterance and ways of understanding what is said. Travis's examples have an unmistakably Justinian flavour that may delight certain readers. Here is a selection of them: 'I present you with this gold watch' (2006, 127). Does the truth of such a statement require all of its parts to be made of gold? Or 'The shoes are under the bed' (2006, 120): is the statement made true by shoes that are located beneath the bed but three floors down? 'The cat has mange' (2006, 181). Does the statement still count as true if there are no mites left but the cat's fur is still in a poor state?

After 1930, Wittgenstein never tries of pointing out that the same sentence can make very different assertions, but not for the reasons given by Frege and contemporary Contextualists, but on the ground that the assertions in question have a different *grammar.* In *Remarks on Colour* §131 Wittgenstein wrote:

> A language-game: report on the greater lightness or darkness of bodies. – But now there is a *related* one: state the relationship between the lightness of certain *colours.* ... The form of the propositions is the same in both cases ('X is lighter than Y'). But in the first language-game they are temporal and in the second non-temporal.

That two tokens of the same sentence can make very different assertions under different circumstances is a symptom of a difference in the grammar of the concepts involved. The emphasis is not on the specific circumstances under which a given sentence is uttered, but on the different *types* of language games that can be played with it. In order to tell which (Fregean) thought is conveyed by the sentence 'X is lighter than Y' it is not enough that we understand the individual words and the way they are put together – the underlying structure of comparatives, say. Nor is it informative to be told that 'X is lighter than Y' is true in English iff X is lighter than Y. An instance of Frege's or Tarski's equivalence schema fails to tell us which *proposition* is expressed by the same sentence in the process of making a move in the language game (cf. PI §136). Similarly, instances of disquotational principles such as '"Moses" denotes Moses,' '"Nothung' denotes Nothung," '"N.N." denotes N.N' turn out to be all but trivial.

That a grasp of the meaning of a word that we have encountered on previous occasions may often be *insufficient* to determine how the word in question is meant or applied on a new occasion is agreed on (almost) all hands. This, however, does not imply that there is no such thing as understanding the meaning of a sentence type outside a context of use. Admittedly, the understanding of a sentence type does not issue in the understanding of a context-invariant *proposition,* but it definitely narrows down the choice of eligible candidates.

The remark I quoted above on the different uses of predicates such as 'lighter' and 'darker' is reminiscent of similar remarks that occur in

TLP (cf. 4.123), and in Wittgenstein's writings of the Thirties. Possibly Wittgenstein never completely gave up his belief that there is a *structure* underlying our perception of colour. Following in the footsteps of the tradition initiated by Riemann, he thought that there is a space of colours in the visual field, in a sense in which there is no similar space of odours (i.e., the manifold of odours does not form a structure), and that from the relations that hold at the phenomenological level one may draw inferences concerning the relations that hold at the physical level. Of course, the correspondence is not one to one: there is much in the physical space that has no representative in the visual space. What in the 30s is discussed in terms of structure and manifolds , in 1950 is discussed with reference to the different moves that are appropriate to the same-sounding sentence in the performance of different language games.

Wittgenstein's aim here is to point out that sameness of wording may conceal profound differences in the *grammar* of the concepts involved. In order to appreciate what is implicit in our concept of colour we are invited to survey the grammar of colour words, and the different roles they play in various interlocking language games. Elsewhere (Picardi, 2009) I argued that Wittgenstein's technique of *extending* a language game, by adding new words together with the corresponding moves, was among his favoured tools of conceptual clarification. The point of *imagining* what such extensions may be like is not that of testing occasion sensitivity and meaning pliability, but to uncover how the bits and pieces that make up our conceptual repertoire hang together and connect with forms of life. Whereas at the time of the TLP internal relations were considered the hallmark of the a priori, in Wittgenstein's later philosophy the uncovering of 'grammar' is supposed to provide a clarification of concepts by means of a descriptive survey of the ways we use words that are taken to express such concepts.

As is well known, the idea (l) of conceptual analysis has been subjected to powerful criticisms on the part of Quine, on the ground that it involves the analytic/synthetic distinction, and on the part of Grice, on the ground that it confuses questions belonging to semantics, having to do with what 'a sentence literally and in the strict sense *says*,' with pragmatic issues, relating to what a sentence conventionally and conversationally implicates. Whether such criticisms also apply to Wittgenstein's idea(l) of conceptual clarification is far from clear.[8] One of the chief merits of today's discussions of the semantics/pragmatics divide is to prompt a critical re-examination of both Quine's and Grice's legacy.

3 Proper names in Frege and Wittgenstein (I)

What I propose to do in the rest of this chapter is to focus on the bearing of (CP 1), (CP 2) and (CP 3) on the meaning ('sense' in Frege's terminology) of proper names. Identity statements play a crucial role in Frege's elucidation of what counts as an object. The difficulty of comparing the bearing of (CP 1) and (CP 2) on issues relating to proper names, depends, among other things, on Wittgenstein's understanding of identity and of the identity sign.

For instance, Robert Fogelin has raised the question how it comes that Wittgenstein never evinced an interest in Frege's 'puzzles' concerning the Morning Star and the Evening Star.[9] These are issues over which philosophers have agonized over the last 40 years. Though Kripke's, Kaplan's, Putnam's 'new' theories of reference have mobilized powerful arguments against Frege's supposedly descriptivist account of proper names, there is no consensus on a satisfactory solution to 'Frege's puzzle(s).' In the framework of TLP Frege's 'puzzle' can hardly be framed, let alone answered, and this may be good news to direct reference theorists. Let's see why.

I have alluded already to some main tenets over which there is (qualified) agreement between Frege and young Wittgenstein. With explicit reference to Frege (and Russell), Wittgenstein says that he considers the meaning of sentences a function of their component parts (TLP 3.318), and that to understand a proposition is to know what is the case if it is true (TLP 4.024). The possibility of understanding *new* sentences depends on our being familiar already with their component parts. They also agree that it is in virtue of being logically articulated that a sentence can say something true or false about the world (TLP 3.141): an isolated name or a string of names cannot fulfil such a function. However, Wittgenstein's emphasis on structure and composition is only a step in the process of establishing the Picture Theory. He holds that in order for a proposition to be able to communicate a *situation* to us, it must be *intrinsically* connected with the situation in question. Frege had failed to show how propositions connect with the world, for he had failed to realize that a proposition is a *fact*, and wrongly compared a proposition to a composite name (TLP 3.143). And this is why Frege's explanation of the concept of truth is unsatisfactory. Unlike the author of *The Problems of Philosophy*, Wittgenstein holds that we cannot possibly be presented with objects in isolation, but only with objects as constituents of state of affairs. The ontology of TLP has structure and modality built into it from the very outset. Section 2.0122 of TLP reads as follows:

> Things are independent in so far as they can occur in all *possible* situations, but this form of independence is a form of connection (*Zusammenhang*) with the states of affairs, a form of dependence. (It is impossible for words to appear in two different roles: by themselves and in a proposition.)

Here Wittgenstein suggests, as it were, a metaphysical foundation for Frege's (CP 1). Unlike Frege, however, Wittgenstein urges that we apply the dependent/independent contrast to both objects and concepts (Frege's unsaturated entities). And this has led some interpreters to count properties among Tractarian objects.[10] What seems to speak in favour of this interpretation is that in TLP the word 'name' is not used in Frege's or Russell's sense of 'Eigenname,' that is, as a device of singular reference. As a matter of fact, the expression 'proper name' never occurs in TLP. What binds together the component 'names' into a propositional unity are the modes of signification (Bezeichnungsweise) appropriate to them (3.322).

Elementary propositions are said to provide the proper habitat of names: 'The name occurs in the context of the elementary proposition' (TLP 4.23). Names occur in the context of elementary propositions, much as objects occur in the context of a (possible) state of affairs. Names are said to be representatives of simple objects in elementary propositions. Possibly, only variables can play such a role, and indeed the variable name 'x' is the proper sign for the pseudoconcept object (TLP 4.1271). And much the same holds for the pseudoconcepts number, proposition, fact, function, complex, etc. (TLP 4.1272). In sum, there is no mention of ordinary proper names in the sense of either Frege or Russell, and this, together with Wittgenstein's loose employment of the expression 'name,' may lead one to think that in TLP (CP 2) has no special bearing, or no bearing at all, on proper names. But this would be wrong: for one feature that (CP 1) and (CP 2) have in common is the emphasis on the referential role that names play in the context of a proposition. Names contribute via their referent to the truth conditions of the sentences in which they occur. As I said, the Picture Theory is supposed to show how Tractarian names intrinsically connect to their (logically structured) referents. Fregean names contribute via their sense to the expression of a thought; however, since no account of the mechanism of reference is forthcoming, Frege's suggestion that the sense of a proper name enshrines a way in which the object is given to us must have struck Wittgenstein as utterly idle. In TLP identity statements do not count as pictures of states of affairs, and hence no need was felt to explain how names connect to their referents in the context of identity statements.

The interpretation of identity statements and of the identity sign brings to the fore one the major differences between Frege and Wittgenstein, perhaps THE major difference. According to Frege, one of the privileged sentential contexts in which proper names can occur is provided by identity statements. Identity is not among the Urzeichen of Wittgenstein's logical syntax. Had Frege accepted Wittgenstein's improved conceptual notation, he would have deprived himself of one class of statements, that is, identity statements, on which he had rested two of his major claims: (1) that identity statements can add to our knowledge and (2) that arithmetical equations express thoughts that are analytic, knowable a priori, and informative. Identity statements also help to clarify the peculiar role that proper names play in the context of a sentence, for no other expression can flank the identity sign. However, it is not just in order to account for the cognitive significance of identity statements that Frege (1892) introduced the distinction between sense and reference. It was essential to his picture of meaning that proper names have meaning (a sense), on a par with other words of the language. And, as we shall see, much the same is true of the author of PI.

In Wittgenstein's opinion what look like contingent statements of identity, such as 'a = b,' read right; tell us nothing about the object named by the signs 'a' and 'b.' 'a = b' simply records our intention to use the two signs flanking the identity sign interchangeably. Statements of identity provide a licence for substituting two expressions for one another. The signs themselves are mere representational devices, and say nothing about the meaning of the signs 'a' and 'b' (TLP 4.242). As I said, the only function of the identity sign is to licence the substitution of a name with a typographically different one or of a longer expression with a shorter one, as is the case with definitions (TLP 4.241). TLP 3.317 reminds us that 'the only thing essential to a stipulation is that it is merely a description of symbols and states nothing about what is signified (das Bezeichnete).' In an ideal conceptual notation difference of objects should be signalled by difference of signs (TLP 4.241).

For to say of one thing that it is identical with itself is a statement that is never worth making, and to say of two things that they are identical is obvious nonsense (TLP 5.5303). In asserting an identity statement we are on the brink of speaking nonsense in so far as we represent ourselves as conveying information about the world, and not about the way signs belonging to the logical syntax ought to be employed. But the use of a symbol belonging to logic syntax shows itself in the application that we make of it. This is why all propositions in which the identity sign occurs turn out to be pseudopropositions (TLP 5.534). And this also explains why there can be no contingent identity statements in Frege's sense. Or so it seems.

As I said, in the framework of TLP Frege's 'puzzle' can hardly be framed, let alone answered. The only section of TLP where Wittgenstein asks a question akin to Frege's 'puzzle,' as applied to contingent statements of identity, is 4.243. There he wonders: 'Can we understand two names without knowing whether they designate (bezeichnen) the same thing of two different things? Can we understand a proposition in which two names occur without knowing whether they mean the same or something different (ob sie Dasselbe oder Verschiedenes bedeuten)?.'[11] But Wittgenstein fails to answer the question. For he changes the subject, and asks a different question, namely whether it is possible to know the meaning (*bedeutung*) of an English word and of a German word without realizing that they are synonymous (*gleichbedeutend*). He apparently takes it for granted that if one understands the meaning of isolated words, she must also be aware that they coincide in meaning; for he goes on to say that one must be able to translate each word into the other.[12]

To the best of my knowledge, Frege's puzzle of the Morning Star and the Evening Star is mentioned by Wittgenstein only in the Big Typscript and in his conversation with O.K. Bouwsma (1986) that took place during the years 1949–51. A puzzling feature of proper names is that one may understand them, in the sense of being able to use them in a variety of sentential contexts, without realizing that one is referring to the same thing. Possibly Wittgenstein thought that such a situation can hardly arise with Tractarian names. If it arises, it only shows that we are using the names as descriptions, for one can well understand two descriptions without realizing that they actually apply to the same bearer. In the *Big Typescript* Wittgenstein seems to favour a meta-linguistic reading of the significance of proper names, for the meaning of 'N' is said to be equivalent to that of the meta-linguistic description 'the bearer of "N"'. The significance of an identity statement would then boil down to the piece of information that a given object is the bearer of two different names. This, at any rate, is the interpretation I am inclined to place on the only occurrence of Frege's example of the Morning Star and the Evening Star that I have come across in Wittgenstein's writings (cf. BT, p. 27).

But what about 'necessary' identity statements? Also concerning mathematical equations Wittgenstein sharply disagrees with Frege. It is instructive to read some of these sections in full, for this is the only place of TLP where Frege's distinction between sense and reference is explicitly mentioned (and its application discarded). With reference to the expressions '1 + 1 + 1 + 1' and (1 + 1) + (1 + 1) Wittgenstein says:

> (6.232) Frege says that the two expressions have the same meaning but different senses. But the essential point about an equation is that it is not necessary in order to show that the two expressions connected by the sign of equality have the same meaning, since this can be seen from the two expressions themselves.
>
> (6.2322) It is impossible to *assert* the identity of meaning of two expressions. For in order to be able to assert anything about their meaning, I must know their meaning, and I cannot know their meaning without knowing whether what they mean is the same or different.
>
> (6.2323) An equation merely marks the point of view from which I consider the two expressions: it marks their equivalence in meaning (*Bedeutungsgleichheit*).

I cannot possibly enter the discussion of these tantalizingly difficult sections of TLP. On the most plausible interpretation I can think of, the word 'meaning' ('Bedeutung') means meaning, significance, and not reference. Yet, it is not completely clear whether, in formulating Frege's conception of identity statements, Wittgenstein consistently uses 'Bedeutung' in the sense of 'significance' (cf. TLP 4.243). Although Wittgenstein himself states, in agreement with Frege, that the object is the *Bedeutung* of a name in the context of a proposition, he holds that in the case of mathematical equations all mention of reference drops out as irrelevant. In mathematics we are interested in internal relations. Wittgenstein here invokes the idea of a series of forms. In the case of the natural numbers, it is a series generated by an internal relation corresponding to the successor relation. The theory of formal series constitutes Wittgenstein's own account of the significance of mathematical statements (cf. TLP 6.24 – 6.241).

So far the result of our investigation is somewhat frustrating. On the one hand, Wittgenstein's thesis that a name is a representative of an object in a proposition suggests a considerable degree of agreement with

Frege. Among other things, it brings to the fore that a name contributes via its referent to the truth conditions of the sentence in which it occurs. Once the notions of truth and truth conditions cease to play a key role in Wittgenstein's reflections on meaning, *this* specific application of Frege's Context Principle is no longer available. And this is what makes the interpretation of PI §49 so puzzling. On the other hand, if we focus on Wittgenstein's treatment of identity statements in general and of mathematical equations in particular, differences clearly outweigh similarities. Following Quine's lead we have got so accustomed to connecting issues of naming with issues of identity, quantification, and ontological commitment, that Wittgenstein's approach to identity strikes us as very contrived.[13]

4 Proper names in Frege and Wittgenstein (II)

What I would tentatively like to suggest in the final part of my paper is that as far as proper names are concerned, the similarities between Frege and the later Wittgenstein by far outweigh the differences. A good starting point is to read (CP 1) in its original setting. In §62 of *Foundations of Arithmetic* Frege writes:

> How then is a number given to us, if we cannot have any idea or intuition of it?
>
> Only in the context of a proposition do words mean something. It will therefore depend on defining a sense of a proposition in which a number word occurs. As this stands, this leaves much undetermined. But we have already established that number words had to be understood as standing for independent objects. This gives us a class of propositions that must have a sense – propositions that express recognition of a number as the same again. If the symbol *a* is to designate an object for us, then we must have a criterion that decides in all cases whether *b* is the same as *a*, even if it is not always in our power to apply this criterion.

Frege's insight that the expression 'the same F,' where 'F' is a sortal concept,[14] requires a criterion of identity if a statement of identity is to be of any use to us, was lost on the author of TLP. However, it is undeniable that after 1930 Wittgenstein came to appreciate the importance of Frege's insight,[15] while refraining from Frege's Platonist construal of it. *We* may not be able to apply a given criterion of identity under all circumstances, but, in Frege's opinion, this does not impugn the soundness of the criterion. Even a cursory reading of Wittgenstein's PI shows that the requirement that a criterion of identity be available, if the expression 'the same F' is to have a point, was to become one of the leading themes of his philosophy. To claim that 'N. N. is in pain' *means* that things are with N. N. as they are with me when I am in pain, seems to offer an explanation of the way we use words for sensations. But it is at best uninformative, and at worst misleading. This and related claims have proved to be among the most controversial aspects of Wittgenstein's philosophical legacy. After 1891 Frege, while mentioning statements of identity and judgements of recognition, never paused to discuss the sort of criteria we rely upon in *recognizing* something as the same planet, the same chemical substance, the same man, that we have come across on a previous occasion. Identity and identification tacitly fell apart, and he treated the identity sign as indicating a relation that either holds or fails to hold between objects, no matter whether we have means to find out whether the expressions flanking the identity sign stand for the same object or fail to do so.

Many philosophers, such as for example, Baker and Hacker (1980), Kenny (1995), and, more recently, Howard Wettstein (2004), have construed Wittgenstein's later philosophy as a reaction against Frege. Wettstein (2004) suggests that the aim of Wittgestein's later philosophy was to oppose the 'Cartesian' picture of mind and language he had inherited from Frege. 'Linguistic contact without cognitive contact' is the motto of those who guided the revolution against the Linguistic Cartesianism that he sees embodied in Frege's philosophy.[16] Among those who first reacted against Linguistic Cartesianism, thus foreshadowing the revolution carried out by Kaplan, Kripke, Putnam, and Donnellan, Wettstein counts the author of PI. In particular, Wittgenstein's conception of proper names is akin in spirit to the social theory of reference advertized by Putnam under the slogan 'Meaning ain't in the head.' Now, there is no denying that the emphasis on the social practice of speaking a language is an aspect that Wittgestein's conception of understanding shares with Putnam's. And yet, it seems to me that Wettstein definitely goes too far in playing Wittgenstein against Frege, and placing him in the camp of the Revolutionaries. However, Wettstein can't possibly ignore the fact that in PI Wittgenstein shows a keen interest in the way we use and *understand* proper names. For one thing, the author of PI, unlike the author of TLP, evinces little interest in the *mechanism* of reference. In Wittgenstein's survey of the ways we use proper names, the referential function no longer plays the pivotal role that was allotted to Tractarian names. To be sure, proper names perform many different functions on which the Revolutionaries have laid great emphasis. For one thing, proper names are capable of fulfilling a role also outside the context of a sentence. They can be employed as labels (cf. PI §§15, 26), they can be used to call people (PI §27), and sometimes the meaning of a proper name can be explained by pointing to its bearer (PI §43).

However, in PI Wittgenstein's main suggestion seems to be that, contrary to what Frege held, there is no *uniform* account of the semantic role that expressions belonging to the grammatical category of proper names play in the language. What is required is a survey of our name-using practices. As a result of such survey we may find out that there is no uniform semantic treatment encompassing all the items that fall under the grammatical category of a proper name. After the rejection of the philosophy of logical atomism, 'vacuous' names such a 'Nothung,' proper names of undistinguished people such as 'N.N.' and proper names of very significant figures such as 'Moses' are allowed to reenter the philosophical scene. (CP 3) has to do both with naming and with names, and important sections of PI are devoted to the meaning and use of proper names such as 'Moses' (PI §79), 'Nothung' (PI §44), 'N' (PI §§38, 41, 42, 45) and to the differences and similarities between the sentential contexts (*Satzzusammenhang*) in which proper names and demonstrative pronouns are appropriate (PI §38)

At least, this is what I take to be the gist of Wittgenstein's remarks in PI concerning proper names. On the face of it, this looks like a substantial departure from Frege's conception of how proper names function in the context of a sentence. And yet, Wittgenstein's way of approaching the issue of the semantic role of proper names in PI is not antagonistic to Frege's. There is room for comparison: for now the attention is focussed on the *understanding* of the use of proper names.

The concept of a proper name is one of the first family-resemblance concepts that we encounter in PI. At any rate, this is how I propose that we interpret Wittgenstein PI §79. That a name like 'Moses' (actually, most names) is used without a fixed meaning does not render the name useless or defective. As Baker and Hacker (1980, p. 425ff) point out, Wittgenstein contention that we use the name 'Moses' without a fixed meaning (*ohne feste Bedeutung*) may be taken to mean either that the name 'N' is used with a family of meanings or that there is a set of meanings among which it fluctuates. Correctly, too, they point out that according to Wittgenstein 'there is no such thing as the name-relation. Rather there are many different relations of names to objects' (Baker and Hacker, 1980, p. 244). Baker and Hacker, however, are clearly dissatisfied with Wittgenstein's way of treating proper names on a par with other words of the language. Wittgenstein's attitude seems to them too close to Frege's and Russell's supposedly descriptivist approach. I tend to agree with them, but, unlike them, I do not find any fault with this similarity, provided that we correctly understand the role played by the descriptive ingredient. Peter Geach made a related point in 1967, in a paper entitled 'The Perils of Pauline':

> A proper name can never be logically tied to just one definite description; so long as we agree in a good many of the judgements we make using a certain proper name, we can use that name for communication. And there is no one judgement mentioning a peculiarity of Pauline such that agreement on *it* is indispensable. ... Frege called mathematical signs like 'π' proper names: and in this respect 'π' really is like a proper name: for π can be joined in an equation to any number or complex mathematical expressions, and is not shorthand for one of these expressions rather than another (Geach, 1972, p. 156).[17]

To be sure, there are different things to be said about the way in which the sense of a proper name contributes to the determination of its referent in the context of a sentence, on the one hand, and to the *communication* of the thought expressed by the sentence in question on a given episode of conversation, on the other. Frege believed that what needs to be shared for communication to succeed is the very same ingredient of the sense a proper name that contributes to the determination of its referent. This aspect of Frege's doctrines has proved very controversial on several grounds. These tensions come to the fore both in the case of empty names (be they fictional, mythological, or historical) and in the case of names of ordinary people, where there is no background of shared traditions to anchor their use, but only a loose and fluctuating set of information in the repertoire of the people directly or indirectly acquainted with their bearers.

If we look more closely at Frege's writings, we realize that he was ready to admit that different things can be said about the sense of proper names belonging to different categories. In the case of proper names of ordinary people what matters are the (fluctuating) pieces of *information* that can help speaker and hearer to find out whom they are talking about on a given occasion by using the same proper name. This is at least how I suggest that we interpret Frege's (1918) remarks concerning the name 'Gustav Lauben,' as used by different speakers (including Gustav Lauben himself by means of the pronoun 'I') to talk on what turns out to be the same fact. Frege (1918) argues that in order for a hearer to grasp what thought a given speaker wants to convey on a given occasion, it is not enough that the former be able to pick out from the surrounding scene information which may conceivably bear on the content and truth conditions of what is said. Unlike Russellian propositions, Fregean thoughts are not made up of objects and propositional functions. However, Frege tells us very little on the mechanism whereby perceptually gathered information can be incorporated into the semantic content conveyed by a sentence with indexicals. Indeed, to provide a tenable account of such matters has proved a formidably difficult task.[18] Moreover, leaving indexicals aside, Frege points out that for communication to succeed it is not enough that on a given occasion speaker and hearer refer to the same man when using a proper name in the process of ascribing the same property to him. They must also be aware that they are doing so. Perhaps this goal is achieved when they think of the man in question under the same mode of presentation. In the case of proper names of ordinary people, such as Herr (Doktor) Gustav Lauben, mentioned in Frege's discussion,[19] the possibility suggests itself to interpret the notion of the sense of a proper name appealed to by Frege in terms of the background information that is *shared* by both speaker and hearer on a certain occasion of conversational exchange. It is this piece of shared information, that may fluctuate from case to case, that allows them to know whether they are talking about the same person in using the same proper name. Thus, if I am right, context sensitivity creeps into the use we make of proper names in communication.

Perhaps direct reference theorists can combine Frege's and young Wittgenstein's conceptions of proper names. The semantic content of identity statements such as 'Hesperus is the same as Phosphorus' may be given along Tractarian lines,[20] whereas the assertoric content which is conveyed on a given *communicative* exchange is accounted for by adverting to the pieces of information that are pragmatically salient on the given occasion. These pieces of information may be provided, as Soames (2002) has suggested, by descriptively enhanced names, or partially descriptive names. There are also obvious parallels between Soames's example and Frege's examples. Those who are familiar with Scott Soames's discussion of the examples 'Peter Hempel lived on Lake Lane in Princeton' and 'Carl Hempel lived on Lake Lane in Princeton' will immediately see the parallel (Soames, 2002, pp. 225–40). Soames does not mention Frege's discussion of a similar example. Unlike Frege, who points out that sentences such as 'I have been wounded,' 'Dr. Gustav Lauben has been wounded,' and 'Gustav Lauben has been wounded' may fail to communicate the same information and *hence* express different thoughts, Soames argues that the proposition *expressed* by 'Peter Hempel lived on Lake Lane in Princeton' and 'Carl Hempel lived on Lake Lane in Princeton' is the Russellian proposition containing the bearer of the name and the property indicated by the predicate. However, the *assertive* content of the two sentences may differ, for in communication we rely on a background of shared assumptions. In discussing belief ascriptions Soames (2002) exploits this distinction in order to account for the intuitive plausibility of saying of Harry, who has heard of a famous Professor Hempel active at Princeton University, but is acquainted with Mr. Hempel, the charming old gentleman that he has met many times at the coffee house at Lake Lane, and whom his wife addresses as 'Peter,' that he does not believe that Peter Hempel is Professor Carl Hempel, though he knows full well that Peter Hempel is Peter Hempel. Whether the truth bearer is the proposition expressed or the (descriptively enriched) proposition conveyed by the sentences in question on a given episode of conversation is the object of much debate between Contextualists and Semantic Minimalists.

But there are also names that have a relatively stable set of descriptive material associated with them, and when need arises, we can draw on it to explain to others what we mean by a given name. Possibly, there are names that can be explained *only* with reference to the larger unit of discourse (be it a mathematical theory or a piece of fictional narrative) to which they belong. According to Frege, the best way to come to grips with the number 2 is by attending to the way in which number words contribute to the truth conditions of sentences belonging to this conceptual notation – short of paradoxes. Fictional names such as Odysseus or Nausicaa, too, are given to us in the context of a narrative. However, whereas in the case of arithmetic we have means of stating (and, with luck, settling) the truth conditions of the sentences in which number words occur (above all, identity statements), in the case of fictional names the issue of truth or falsity does not arise. Frege's reasons for encompassing empty proper names in his semantics are twofold: (a) they may occur in *oratio obliqua* clauses, to whose semantic value they contribute via their sense; and, (b), they may occur in quotation marks in sentences such as 'The name "Vulcan" lacks a bearer,' which is one of Frege's two ways of saying that Vulcan does not exist. The other is to incorporate 'Vulcan' into the predicate 'x is-identical-to-Vulcan,' and proceed to explain that to the concept thus indicated belongs the number 0.

There are no fictional objects for empty names to be names of, and therefore we cannot characterize the sense of a proper name as the way in which its referent is given to us. And yet, we are under the impression of understanding the sense of such names. There is a very interesting passage in Frege's posthumously published writings (Frege, 1979, p. 122) concerning the significance of fictional names that may be worth quoting in full:

> It must be determinate for every object whether it falls under a concept or not; a concept word which does not meet this requirement on its meaning (*Bedeutung*) is meaningless (*bedeutungslos*). E.g. the word μῶλυ (Homer, *Odyssey*, X, 305) belongs to this class, although it is true that certain characteristic marks are supplied. For this reason the context cited need not lack a sense, any more than other contexts in which the name 'Nausicaa,' which probably does not stand for nor names anything, occurs. But it behaves *as if* it names a girl, and it is thus assured of a sense. And for fiction the sense is enough. The thought, though it is devoid of meaning, of truth value, is enough, but not for science (my emphasis).

The author of the Odyssey invites his conniving reader to engage in a game of make believe: proper names are consciously used 'as if' they stood for a referent to which genuine predicates apply.[21] We apprehend many singular statements in the context of which the proper name 'Nausicaa' occurs, thereby forming a conception of its significance. Whereas in PI Frege's conception of vague predicates is subjected to a devastating criticism, it is conceivable that Wittgenstein would have raised no objection to Frege's surmise as regards the way in which a name like 'Nausicaa' secures a sense for itself in the context of a narrative.[22] In PI §§40, 41 Wittgenstein himself makes a suggestion along similar lines.[23]

If what I have said is roughly on the right track, Frege's and Wittgenstein's description of the way we use proper names is not rendered obsolete by the picture sketched by causal theories of reference. Moreover, Putnam's slogan 'Meaning ain't' in the head' admits of a variety of construals. Possibly, what enables us to take part in language games is not explicit knowledge of shared meanings, rules and conventions. But it cannot be *ignorance* of such matters, either. Meaning is what the explanation of meaning explains. And such an explanation is not exhausted by adverting to patterns of deference, convergence upon assent and dissent, blind drill, and regularities of conduct. This is a caricature of language as a social institution, and a dangerous one, to boot.

Notes

1. Cf. Frege (1884), §46, §60, §62, §106, and Preface, XXII.
2. I have discussed such matters elsewhere; cf. (Picardi, 2002) and (Picardi, 2007).
3. There is no explicit mention of Frege (1884) in TLP. At that time Frege had not formulated yet the sense/reference distinction. However, the thesis that numbers are logical objects is clearly stated, together with the suggestion that the adjectival role that number words play in the context of a sentence can always be turned into a substantival role, which is truer to the essence of numbers.
4. Many sections of PI are devoted to the criticism of these twin assumptions.
5. As Reck (2001) has pointed out, at all stages of his philosophical career Wittgenstein concurred with Frege on these general tenets.
6. Cf. Dummett (1981).
7. Cf. Dummett (1981).
8. Perhaps what is needed is a shift of emphasis. Under this respect *On Certainty* opens up new perspectives. One of the issues Wittgenstein raises there is not so much what counts as a grammatical proposition vis-à-vis an experiential proposition, but what counts as *being treated as one* within a certain epistemic context. For a survey of the different interpretations to which *On Certainty* lends itself cf. the essays collected in Moyal-Sharrock and Brenner (2007).
9. Both in conversation and in a paper delivered in March 2006 at the University of Bologna.
10. For a survey of various interpretations cf. Frascolla (2007).
11. This is my own translation of 4.243.
12. Readers conversant with the predicament of Kripke's Pierre are vividly aware of the tangled issues involved in inter-subjective and intra-subjective synonymy. Cf. (Kripke, 1979). Moreover, there is no guarantee that one should be able to translate the words in question in all sentential contexts in which they may occur.
13. Wittgenstein's account of identity, as applied to mathematical equations, is somewhat reminiscent of Frege's construal of identity at the time of his Begriffsschrift of 1879.
14. In Strawson's characterization, a sortal (universal) supplies a principle for distinguishing and counting individual particulars which it collects (1959, p. 168).
15. Possibly he had come to appreciate this point independently of Frege. My surmise that the reading of Frege's Foundations of Arithmetic might have played a role in this connection stands in need of textual evidence.
16. Wettstein 2004, p. 75.
17. Geach's characterization of the significance of proper names in the language is worth recalling:

    I do indeed think that for the use of a word as a proper name there must be someone acquainted with the object named. But language is an institution, a tradition; and the use of a given name for a given object, like other features of language, can be handed down from one generation to another; the acquaintance required may be mediate, not immediate.

> Plato knew Socrates, and Aristotle knew Plato, and Theophrastus knew Aristotle, and so on in apostolic succession down to our times; that is why we can legitimately use "Socrates" as a name as we do. It is not our knowledge of this chain that validates use, but the existence of such a chain [...] When a serious doubt arises (as happens for the well known use of the word 'Arthur') whether the chain does reach up to the object named, our right to use the name is questionable, just on that account. But a right may obtain even when it is open to question. (Geach, 1972, p. 155)

Geach's appeal to a chain of communication stresses the *social* dimension of the use of proper names; this feature of our use of proper names is also stressed by Kripke (1980).

18. Suffice it to think of the work done in this area by David Kaplan, Hans Kamp, John Perry and Gareth Evans.
19. The (Herr Doktor) Gustav Lauben mentioned in Frege (1918) was an acquaintance of Frege. I owe this piece information to Gottfried Gabriel.
20. Provided that identity is given a more robust metaphysical construal.
21. I am here thinking of Kendall Walton's work on make-believe (Walton, 1990). Cf. also Evans's discussion of make-believe (Evans, 1982, ch. 10), and Carpintero (2006). For a survey of the vast literature on fictional names cfr. Sainsbury (2005) and Voltolini (2006). – In reading Frege's text one is reminded of Hans Vaihinger's Philosophie des Als Ob. To be sure, Frege would not have appreciated Vaihinger's version of fictionalism as applied to scientific discourse. Moreover, Vaihinger's book was first published in 1911, and Frege could not possibly have alluded to it in his 'Comments on Sense and Meaning' that were probably written between 1895 and 1899. However, Vaihinger's interpretation of Kant's philosophy was widely known before that date.
22. Plainly this account does not generalize – nor would we want it to generalize – to all empty names. The case of the names 'Vulcan' and 'Neptune' provides us with an excellent example. Le Verrier put forward the hypothesis that two planets, for which he provided proper names and rigidified descriptions, caused certain anomalies in the orbits of Mercury and Uranus, respectively. His conjecture turned out to be correct in the case of Uranus, but failed in the case of Mercury. As a matter of fact, the explanation of the strange appearance of Mercury's orbit was provided much later, in the framework of Einstein's theory of relativity. Plainly, Le Verrier did not engage in a game of make-believe, but in the practice of scientific explanation. Indeed, we want to say that Le Verrier was right in one instance and wrong in the other, not that he had slipped inadvertently into fiction, as Frege intimates. The evidence on which he based his conjecture, together with his mathematical calculations, proved right in the case of Neptune, but inadequate in the case of Vulcan. However, for all that was known at the time, there might have been a planet between Mercury and Venus. Or so it seems to me. I find the idea that everything that does not exist, does not exist of necessity, very difficult to support on the basis of our present understanding of what counts as a planet of our solar system.
23. I discuss these sections in Picardi (2009).

9
Sceptical Conclusions about Epistemological Analyticity*

Timothy Williamson

Among the topics of discussion to which Diego Marconi has contributed outstandingly over the years, two of the most notable are the nature of lexical competence and the status of the law of noncontradiction. The topics are linked by the popular idea that adherence to basic logical laws is a necessary condition of understanding logical words, in particular, that adherence to the law of noncontradiction is a necessary condition of understanding words for negation. For example, it may be proposed that someone who asserts both 'It is raining' and 'It is not raining' in the very same context thereby demonstrates semantic incompetence with the word 'not.' This is to attribute a sort of *epistemological analyticity* to a form of the law of noncontradiction. Diego Marconi's work contains many important correctives to this crude picture, with respect to both the nature of lexical competence and the status of the law of noncontradiction. The present chapter aims to clarify the prospects for some form of epistemological analyticity, with particular reference to the relation – or lack of it – between lexical competence and adherence to basic logical laws. Although it grants the intelligibility of a notion of linguistic meaning, it still reaches sceptical conclusions about the supposed phenomenon of epistemological analyticity.

1

If someone is unwilling to assent to the sentence 'Every vixen is a female fox,' the obvious hypothesis is that they do not understand it, perhaps because they do not understand the word 'vixen'. The central idea behind epistemological conceptions of analyticity is that, in such cases, failure to assent is not merely *good evidence* of failure to understand; it is *constitutive* of such failure. The unqualified link from understanding to assent is this:

UA Necessarily, whoever understands the sentence 'Every vixen is a female fox' assents to it.

One proposal is to generalise UA to define an epistemological notion of analyticity: a sentence *s* is analytic just in case, necessarily, whoever understands *s* assents to *s*.

Four glosses on UA must be taken as read throughout. First, it concerns 'Every vixen is a female fox' with its current meaning, for of course if the phonetically individuated sentence had meant something different, someone might easily have understood it and refused to assent. Second, assent is dispositional, for of course we are not actively assenting to any sentence whenever we understand it. Third, assent is a mental attitude, not a merely verbal one, for someone might easily understand 'Every vixen is a female fox' while refusing to give it overt assent, for example because overt assent to a triviality looks uncool. We could speak of belief rather than assent, but the latter term sounds more natural in relation to inference rules, to which the notion of analyticity is generalised. Fourth, assent is not a metalinguistic attitude: normally, in assenting to 'Grass is green,' one is in effect saying or thinking that grass is green, not that the sentence 'Grass is green' or the thought *grass is green* is true. Assenting to the sentence 'Grass is green', for someone who understands it, is something like believing that grass is green under the guise of that sentence. In a context in which the sentence *s* expresses the proposition *p*, assenting to *s*, for someone who understands it, is something like believing *p* under the guise of *s*. For you, assenting to 'I am hungry' is something like believing that you are hungry under the guise of the sentence 'I am hungry,' since in your context that sentence expresses the proposition that you are hungry, not the proposition that I am hungry.

The notion of an understanding-assent link generalises from individual sentences to rules of inference. For example, if someone is unwilling to assent to the inference from 'This is red and round' to 'This is red,' the obvious hypothesis is that they do not understand one of the sentences, most probably because they do not understand the word 'and.' For epistemological conceptions of analyticity, failure to assent in such cases is again not merely good evidence of failure to understand but constitutive of such failure. Gerhard Gentzen introduced the idea that some rules of his natural deduction systems of logic have definitional status. Following him, a tradition which includes Dag Prawitz, Michael

Dummett, Per Martin-Löf, Christopher Peacocke, Robert Brandom, Paul Boghossian and many others has developed in various ways the conception of acceptance of such inference rules as playing a constitutive role in understanding the logical constants, and therefore in understanding the sentences in which they occur. For many of these thinkers, this is one step towards a quite general 'inferentialist' account of meaning and understanding for expressions in terms of their conceptual roles.[1]

Understanding-assent links, or something like them, are commonly thought to play a leading role in the understanding of theoretical terms in science: if you don't assent to some core sentences of electron theory, in which the word 'electron' occurs, you don't understand the word, and therefore don't understand those sentences.

Given an understanding-assent link, what are its epistemological consequences? Our concern is knowledge or justification, not just belief or assent. Ideally, we want understanding-knowledge links like these:

> UK Necessarily, whoever understands the sentence 'Every vixen is a female fox' knows 'Every vixen is a female fox.'

Here, knowing 'Every vixen is a female fox' amounts to knowing that every vixen is a female fox under the guise of the sentence 'Every vixen is a female fox.' Since knowing something entails assenting to it (we may assume), UK entails UA. But since assenting to something does not entail knowing it, how are understanding-knowledge links to be extracted from understanding-assent links? UA does not entail UK in any obvious way.

An even more elementary problem arises. Knowledge is factive. Thus understanding-knowledge links entail corresponding understanding-truth links:

> UT Necessarily, someone understands the sentence 'Every vixen is a female fox' only if it is true.

Thus if understanding-assent links somehow imply the corresponding understanding-knowledge links, *a fortiori* they also imply the understanding-truth links. Perhaps UT holds because the sentence 'Every vixen is a female fox' is necessarily true. But in other cases the question of truth becomes more urgent.

Consider theoretical terms from discredited theories. If an understanding-assent link holds for 'phlogiston', and understanding 'phlogiston' necessitates assent to a core of phlogiston theory, how could it follow that someone understands sentences of phlogiston theory only if a core of it is true? Didn't proponents of phlogiston theory understand their own theory, despite its actual untruth? Likewise, people who advise us to drop various ordinary terms, on the grounds that obsolete folk theories are built into the use of them, may be assuming that while an understanding-assent link obtains for critical sentences of the folk theory, the corresponding understanding truth-link fails (if so, they presumably do not count themselves as fully understanding the folk theory). For if one could understand the sentences which make up the false theory without assenting to them, in what sense is the theory built into the use of those terms? For example, we could use the terms to assert the negations of central principles of the theory.[2]

Some understanding-assent links might even be to logically inconsistent sentences. For example, the ordinary notion of truth is sometimes held to be incoherent, on the grounds that a necessary condition for understanding 'true,' and so for understanding sentences in which it occurs, is assent to a disquotational principle for 'true' which the Liar paradox shows to be inconsistent. Tarski's description of natural languages as 'inconsistent' in virtue of the paradox (1983, pp. 164–5) may involve such a view, for if we can understand 'true' in English without assenting to the troublesome instances of the disquotational principle, what prevents us from using English consistently?[3] Similarly, Prior's connective 'tonk' has mismatched introduction and elimination rules with which any conclusion can be derived from any premise; the introduction rule licenses the inference from 'P' to 'P tonk Q,' while the elimination rule licenses the inference from 'P tonk Q' to 'Q' (Prior, 1960). If assent to instances of those rules is necessary for understanding them, because necessary for understanding 'tonk', it hardly follows that the rules are truth-preserving (in the context of someone who understands 'tonk'); they are so only if either every sentence or no sentence of the language is true (including atomic sentences, in which 'tonk' does not occur).[4]

Such examples can be interpreted in diverse ways. Nevertheless, they show at least that to advance from understanding-assent links to understanding-truth links, let alone to understanding-knowledge links, is no trivial task.

One response to the examples is to stop trying to link understanding to knowledge and truth in this way, and try only to establish links to justification, conceived as nonfactive. The hope would be to reach understanding-justification links like this:

UJ Necessarily, whoever understands the sentence 'Every vixen is a female fox' is justified in assenting to it.

But this retreat from knowledge and truth to justification does less than full justice to the examples. Imagine a dogmatic proponent of phlogiston theory, who continues to accept it long after the accumulating negative evidence has made this unjustifiable. Suppose that 'phlogiston' does indeed provide a counterexample to the putative entailment from the understanding-assent link to the understanding-truth links. Thus although understanding a core of phlogiston theory necessitates assent to that core, because understanding the core necessitates understanding the term 'phlogiston' and understanding 'phlogiston' necessitates assent to the core of phlogiston theory, someone can understand the core despite its untruth. But if anyone can understand the core of phlogiston theory, its proponents can. Moreover, they do not stop understanding it when they unjustifiably refuse to take seriously the mounting negative evidence. Thus our last-ditch defender of phlogiston theory understands its core but is unjustified in assenting to it: the understanding-justification links fail too. For more blatantly defective concepts, the assent mandated by understanding-assent links may be unjustifiable from the start, as with 'tonk'. In such cases too, an understanding-assent link which lacks the understanding-truth link also lacks the understanding-justification link. The examples do not motivate a retreat from knowledge and truth to nonfactive justification. Rather, if they work, they show that some understanding-assent links have no positive epistemological upshot.

A different response to the examples is that they do not work: either the understanding-assent link fails or the understanding-truth link holds.

Since the relevant sentences in the examples are clearly untrue, the understanding-truth link can hold in them only vacuously. That is, in such pathological cases, understanding is impossible: no meaning is there to be grasped.[5] This response seems plausible for 'tonk'. It also makes the links from understanding to truth and any positive epistemic status hold vacuously. Where there is no understanding, we can hardly expect much of a positive epistemological upshot from a constraint on understanding. A trickier question is whether such cases of an illusion of understanding have negative epistemological repercussions for cases of genuine understanding, since a sceptical doubt can arise for the subject in the latter cases too as to whether the understanding is not an illusion. If it could avoid such repercussions, this response might maintain a general entailment from understanding-assent links to understanding-knowledge links and the rest. However, the response is less plausible for 'phlogiston', 'true' and some of the other examples than for 'tonk'.[6] There is some difference between understanding the word 'phlogiston' or 'true' and not understanding it. Even if sentences with 'phlogiston' or 'true' fail to express propositions, because 'phlogiston' or 'true' fails to refer, they may still have linguistic meanings. Demonstratives such as 'this' have linguistic meanings, teachable to non-native speakers even in contexts in which no attempt has been made to assign them a reference. If a meaningful expression like 'this' can lack reference in some contexts, perhaps a meaningful expressions like 'phlogiston' can lack reference in all contexts.

Alternatively, someone might maintain that the understanding-assent links in these examples fail, but that understanding-link for other sentences hold. Thus the examples involve genuine understanding. On this view, understanding-assent links may still be held to entail the corresponding understanding-knowledge links. It claims that the examples selected the wrong candidates for understanding-assent links. Either such links hold only for nondefective words or for those defective cases they hold only for cautiously circumscribed sentences. For instance, rather than the core of phlogiston theory itself, we might have the conditional 'If phlogiston exists then ...,' with that core filling in the dots. Arguably, however, since 'phlogiston' fails to refer, that conditional too fails to express a proposition, so even this more cautious sentence is not true, although it is also not false. A more general objection is that this response treats our practices as though they are bound to have anticipated from the beginning all problems which could subsequently arise for them. Presumably, if understanding-assent links hold, they do so because they are part of the linguistic practices at issue. Suppose, for instance, that understanding 'true' necessitates assent to a disquotational principle carefully and ingeniously modified to avoid all the semantic paradoxes. But since they scarcely ever arise in ordinary life, why was our ordinary practice with the word 'true' tailored in advance to avoid them? Indeed, the puzzlement they cause suggests quite the opposite. That such precautions are part of every possible linguistic practice is even less likely. If understanding-assent links hold for some other reason than that they are part of the linguistic practices at issue, what is that other reason? Even if one moderates the approach by substituting understanding-justification links for understanding-knowledge links, a version of the objection still applies. If our linguistic practices can make assent to inference rules a precondition of understanding, nothing seems to stop foolish practices from requiring such assent to rules which blatantly generate consequences not involving the original word at issue, as with 'tonk'. Such consequences may include arbitrary pernicious dogmas (such as racist ones) for which no justification is provided. More cautious fallbacks need not even implicitly have been provided; the practice simply breaks down once the dogma is abandoned. So this alternative way of maintaining a general entailment from understanding-assent links to understanding-justification links, let alone understanding-knowledge links, is unpromising. The objections tell equally against the putative understanding-knowledge or understanding-justification links, even if no attempt is made to *derive* them from understanding-assent links.

A more moderate response concedes that defective practices give rise to understanding-assent links without corresponding links to truth or any positive epistemological status, but maintains that understanding-assent links for nondefective practices do yield such links. For instance, one might try to tell a story on which understanding-assent links for nondefective practices constrain the reference of the relevant words so that the sentences in the links come out true (for defective practices, this constraint cannot be met). Under such conditions, understanding-assent links generate understanding-truth links. Thus assent to those sentences (while understanding them) is, completely reliably, assent to truths. One might hope to squeeze understanding-knowledge links out of such reliability considerations, perhaps when enhanced by an argument that the reliability is not completely hidden from the subject. Clearly, much work would be needed to vindicate such a programme.[7]

A lazy alternative simply postulates understanding-knowledge or understanding-justification links for nondefective practices without attempting to derive them from understanding-assent links. But this has little explanatory value. I understand 'Every vixen is a female fox,' and it has some positive epistemic status for me. How does it get that status? How do I know 'Every vixen is a female fox?' Why am I justified in assenting to it? The lazy theorist may try to dismiss the question, saying that it is simply part of our linguistic practice that 'Every vixen is a female fox' has that positive epistemic value for whoever understands it. But the examples of defective practices show that it is not simply up to linguistic practices to distribute positive epistemic status however they please. That the practice is to treat a given sentence as having some positive epistemic status for competent speakers of the language does not imply that it really has that epistemic status for them. Their belief may be untrue and unjustified, however much the practice deems otherwise. Thus the only plausible way to make the relevant practice guarantee the putative link from understanding to the positive epistemic status is by making absence of the epistemic status constitute absence of understanding, just as absence of assent was supposed to do. On this account, whoever does not know 'Every vixen is a female fox' or is not justified in assenting to it *thereby* fails to understand it. But this direction of explanation does not trivialise the positive epistemic status, to which it assigns the role of constituter, not constituted. Thus the lazy theorist cannot simply dismiss the question: how does 'Every vixen is a female fox' gets its positive epistemic for whoever understands it? Positing direct links from understanding to knowledge or justification does not remove the need for substantive epistemology here. Even when the relevant sentence has the positive epistemic status at issue, the reason is not simply that the linguistic practice deems it to be so – which of course is not to say that the practice is irrelevant to its epistemic status. Understanding-assent links are a better place to start, if not to finish.

2

The standard natural deduction rules constitute perhaps the simplest nontrivial case for an inferentialist account of meaning, and favourite candidates for epistemologically analytic inference rules. By the introduction rule for 'and', a conclusion of the form 'A and B' follows from the corresponding premises 'A' and 'B' together. By the elimination rule for 'and', the conclusion 'A' follows from the premise 'A and B,' as does 'B' from the same premise. Are these rules indeed epistemologically analytic, in the sense that assent to such deductions is a necessary condition of understanding the word 'and'?

One must formulate what acceptance of the introduction rule requires with particular care, since the probability of a conjunction may be less than the probability of either disjunct. Iterations of the introduction rule yield the Lottery and Preface paradoxes. Given a lottery known to have at most a million tickets and only one winner, each premise of the form 'Ticket *i* will lose' is overwhelmingly probable, even though their conjunction is known to be false. The author of a book may endorse each individual statement in it, yet admit in the preface that, despite all her efforts, it is bound to contain errors, and on those grounds reject the conjunct of the individual premises. Of course, these paradoxes do not show that the introduction rule fails to preserve truth, although they might be used as grounds for rejecting the rule by a theorist who (mistakenly) used a probabilistic criterion for acceptance. The elimination rule does not suffer from these problems, since the probability of a conjunction is never higher than the probability of any given conjunct.

Let us therefore concentrate on the elimination rule for conjunction, as having the best chance of being nondiscretionary for competent speakers.[8] Imagine a native speaker of English who shows no sign of linguistic deviance, at least until adulthood. He then starts to be worried by the problem of vagueness. Like many philosophers, although not the present author, he comes to the conclusion that vague statements in borderline cases are neither true nor false. Instead, he regards them as undefined in truth-value. He also reflects on the question of how the status of a complex statement such as a conjunction, disjunction, negation or material conditional as true, false or undefined depends on the status of its constituent sentences. He comes to the conclusion that if every constituent sentence is either true or false, then the complex sentence is true or false, as determined by the classical two-valued truth-table for the relevant connective, but that if at least one constituent sentence is undefined in truth-value, then the complex sentence is undefined in truth-value too. His picture is that since the truth or falsity of the complex sentence is defined as a function of the truth or falsity of the constituent sentences, if one of the latter fails to deliver an input, the definition cannot deliver an output, because the function does not have all the required arguments; consequently, the result is undefinedness. Formally, the upshot corresponds to Stephen Kleene's so-called weak three-valued tables (1952, p. 334). Furthermore, our imagined subject regards undefined statements as within the speaker's discretion, and therefore both truth and undefinedness as acceptable (designated) semantic statuses for an assertion: what matters is to avoid falsity. The joint implication of these views is that any conjunction, disjunction, negation or material conditional with a borderline constituent is undefined in truth-value and therefore assertable – of course, on his view, most such sentences should not be uttered, on the pragmatic grounds that they violate the conversational maxim of relevance (Grice, 1989, p. 27). Now suppose that 'A' is simply false while 'B' is borderline. Consequently, on his view, 'B' is undefined, so 'A and B' is also undefined. Thus the corresponding instance of conjunction elimination – 'A and B; therefore A' – has a designated premise and an undesignated conclusion. On these grounds, he accepts the premise and rejects the conclusion of that instance. He regards the inference as invalid. Let us assume that he is in fact mistaken; the standard elimination rule for conjunction is valid after all. If not, we can change the example, describing new characters who are deviant with respect to some inferences that really are elementary logical validities.

If the standard elimination rule for conjunction is epistemologically analytic, in the sense of satisfying an understanding-assent link, then our imagined speaker does not understand the word 'and', at least, not with its customary English sense. Is that plausible?

In principle, our imagined speaker is highly deviant, and not only with respect to conjunction. For example, he regards both 'B' and 'Not B' as assertable in a borderline case. In practice, however, conversational considerations mask most of the deviance. Like many other people with strange views, he adapts himself to society by not insisting on them unless they are strictly relevant. Occasionally he cannot avoid the problem and sounds pedantic, as many academics too, but that hardly constitutes a failure to understand the words at issue. When challenged on his logical deviations, he defends himself fluently. In fact, he has published widely read articles on the issues in leading refereed journals of philosophy, in English. He appears like most philosophers, thoroughly competent in his native language, a bit odd in some of his views.

It might be objected that he appears to be using the word 'and' in its standard English sense because he is really using it in a sense very similar to, but not exactly the same as, the standard one. Indeed, it may be argued, his nonstandard sense was explained above, by reference to the Kleene three-valued semantics with two designated values, on which conjunction elimination is invalid, whereas by hypothesis it is valid on the standard semantics for English. But matters are not so simple. Our imagined speaker is emphatic that he intends his words to be understood as words of our common language, with their standard English senses. He uses 'and' as a word of that public language. He believes that his semantic theory is correct for English as spoken by others, not just by himself, and that if it turned out to be (heaven forbid!) incorrect for English as spoken by others, it would equally turn out to be incorrect for English as spoken by himself. Giving an incorrect theory of the meaning of a word is not the same as using the word with an idiosyncratic sense – linguists who work on the semantics of natural languages often do the former without doing the latter. Our imagined speaker's semantic beliefs about his own use of 'and' may be false, even if he sometimes relies on those beliefs in conscious processes of truth-evaluation. Indeed, we may assume that he does not regard the three-valued weak Kleene table for conjunction as capturing the way in which the meaning of 'and' is presented to him or any other English speaker simply in understanding it, any more than it is so presented by the classical two-valued truth-table. He regards 'and' as a semantically unstructured connective for which a homophonic statement of meaning would be more faithful. For him, the more elaborate articulations are simply convenient records of important logical facts about 'and'. Only in tricky cases does he resort to his nonstandard semantic theory in evaluating nonmetalinguistic claims.

Our imaginary speaker learned English in the normal way. He acquired his nonstandard views as an adult. At least before that, nothing in his use of English suggested semantic deviation. Surely he understood 'and' with its ordinary meaning then. But the process by which he acquired his eccentricity did not involve forgetting his previous semantic understanding. For example, on his present understanding of 'and', he has no difficulty in remembering why he used to assent to it. He was young and foolish then, with a tendency to accept claims on the basis of insufficient reflection. By ordinary standards, he understands 'and' quite well. Although his rejection of conjunction elimination might on first acquaintance give an observer a defeasible reason to deny that he understood 'and', any such reason is defeated by closer observation of him. He genuinely doubts conjunction elimination as we use it. Nor is he a marginal case of understanding: his linguistic competence is far more secure than that of young children or native speakers of other languages who are in the process of learning English. If some participants in a debate have an imperfect linguistic understanding of one of the key words with which it is conducted, they need to have its meaning explained to them before the debate can properly continue. But to stop our logical debate with the imaginary speaker in order to explain to him what the word 'and' means in English would be irrelevant and gratuitously patronising. The understanding which he lacks is logical, is not semantic. His attitude to conjunction elimination manifests only some deviant patterns of belief and belief-formation. Since there clearly could have been, and perhaps are, people such as our imaginary speaker, we have a counterexample to the understanding-assent link for conjunction elimination.[9]

It is arguable that linguistically competent violations of conjunction elimination are actual, not just possible, in the Conjunction Fallacy, a much-studied psychological phenomenon in which subjects assign a higher probability to a conjunction than to one of its conjuncts.[10]

By tweaking the example, we can construct a counterexample to the original understanding-assent link UA, for 'Every vixen is a female fox.'

Imagine a speaker who uses the same weak Kleene three-valued tables, but treats only truth as a designated value. Furthermore, this speaker treats a universal generalisation of the form 'Every F is a G' as true if and only if the conditional 'x is an F $\rightarrow$ x is a G' is true for every value of the variable 'x' in the contextually relevant domain. Thus he treats 'Every vixen is a female fox' as true if and only if the conditional 'x is a vixen $\rightarrow$ x is a female fox' is true for every such value of 'x'. But there are borderline cases for 'vixen', for example when one retraces the maternal line of a fox back to an evolutionary ancestor that was borderline for membership of the current species. For that value of 'x', 'x is a vixen' is undefined on the envisaged account of vagueness ('Every vixen is a female fox' is intended to cover past animals as well as present ones), so, by the weak Kleene table for $\rightarrow$, 'x is a vixen $\rightarrow$ x is a female fox' is also undefined. Consequently, 'Every vixen is a female fox' receives an undesignated value on the account with only one designated value. Thus our second imagined speaker does not assent to 'Every vixen is a female fox.' Nevertheless, he understands it with its ordinary English sense, for reasons similar to those already sketched.

Our imaginary speakers are not so different from actual natives speakers of English who deny that a man who has lived with a partner for several years without getting married is a bachelor, or assert that someone who underwent a sex-change operation after giving birth is a mother without being a female parent.[11] Suppose that they are in fact mistaken; 'bachelor' is synonymous in some sense that entails sameness of intension with 'unmarried man', and 'mother' with 'female parent'. They falsely believe that 'bachelor' is not synonymous in that sense with 'unmarried man' or that 'mother' is not synonymous with 'female parent'. Nevertheless, they fall well within the range of permissible variation for linguistically competent speakers. They are only giving more weight than others to an inclination that most speakers feel in some degree to classify the cases that way. Without regarding them as having spoken parrot-fashion, we report their beliefs using the words 'bachelor' and 'unmarried'. We classify them as believing that some unmarried men are not bachelors and that some mothers are not female parents because we interpret them as having used the words with their normal English meanings, despite their errors. That is how they intend to be interpreted, not as using the words with idiosyncratic senses.[12] If we believe that all unmarried men are bachelors and all mothers are female parents, we therefore classify their beliefs in question as untrue, for the belief that some unmarried men are not bachelors is true if and only if some unmarried men are not bachelors, and the belief that some mothers are not female parents is true if and only if some mothers are not female parents. Given that we correctly interpret them as using the words with their normal English meanings, they understand the words in the relevant sense of 'understand.' Although they are ignorant of some facts about the normal English meanings of the words, such ignorance is quite compatible with linguistic competence, which is why native speakers of English take university courses in the semantics of English. Arguably, their error is not primarily semantic: they have the semantic belief that the word 'bachelor' does not apply to all unmarried men because they have the nonsemantic belief that some unmarried men are not bachelors and the semantic knowledge that 'bachelor' applies only to bachelors; they have the semantic belief that the word 'mother' does not apply only to female parents because they have the nonsemantic belief that some mothers are not female parents and the semantic knowledge that the word 'mother' applies to all mothers.

Such cases also help answer the objection that, in such cases, the awkward subject who consciously denies that P also has unconscious, semantically derived knowledge (or belief) that P. When a competent native speaker denies that every unmarried man is a bachelor, the postulation of unconscious knowledge (or belief) that every unmarried man is a bachelor serves no good explanatory purpose. The speaker tends to apply 'bachelor' to something once they have applied 'unmarried' and 'man' to it, but the tendency is defeasible. Such defeasible connections can be explained without postulation of unconscious belief in a universal generalisation. In such cases, there need be no hint of the cognitive dissonance or tension that one might expect from a direct contradiction between conscious and unconscious beliefs. Given that there is no contradicted unconscious knowledge in these simple cases, it is not clear what better reason there is supposed to be in postulating it for more complex cases either.

To try to save understanding-assent links by restricting them to rational agents would be pointless. By ordinary standards, our imagined speakers are rational agents. Although they fall short of some high standards of rationality, so do most humans. Understanding-assent links that do not apply to most humans would be of limited epistemological interest. The picture was that those who appear to reject sentences that are candidates for epistemological analyticity can be excluded from the discussion because they lack the linguistic competence to engage in it; but we cannot exclude humans who reject such sentences on those grounds if the connection between rejecting them and lacking competence holds only for superhumans, not for humans.

One might try to salvage something from understanding-assent links by weakening the putatively necessary condition for understanding from assent to a disposition to assent, with the thought that our imagined speakers understand 'and' and other logical words only because they have an unmanifested disposition to make the logically orthodox inferences, masked by a superficial overlay of unorthodox theorizing. Space does not permit detailed exploration of that proposal here. However, empirical evidence from the psychology of reasoning tells against the postulation of underlying classical logical dispositions of the required kind: humans have nothing like a deductive logic module isolated from the effects of unorthodox theorizing. The case for treating lack of a disposition to assent to a sentence such as 'Every vixen is a vixen' as lack of linguistic competence depends on the status of the sentence as an elementary truth of deductive logic. But human deductive competence is far more sensitive than linguistic competence to high intelligence and advanced education. Deductive competence is a reflective skill, often painfully acquired and under one's personal control. It is not insulated from one's conscious theorizing. Thus deductive proficiency is not a precondition of linguistic competence (Williamson, 2006, 2008).

No given argument or statement is immune from rejection by a linguistically competent speaker. Quine's epistemological holism in 'Two Dogmas of Empiricism' undermines his later claim about the deviant logician's predicament: 'when he tries to deny the doctrine he only changes the subject' (Quine, 1970, p. 81).

3

Some special cases deserve separate discussion. For example, some metalinguistic sentences look analytic for distinctive reasons. Consider theoretical terms. We can understand the word 'phlogiston' without believing phlogiston theory. Might we do so because we still believe that 'phlogiston' is generally associated with that theory, just as one can understand a natural kind such as 'gorilla' without believing the associated stereotype ('Gorillas are ferocious') because one still believes that 'gorilla' is generally associated with that stereotype? However, such sociolinguistic beliefs are no more immune than logical beliefs from the challenge of theoretical unorthodoxy without change of meaning. If T is any version of phlogiston theory, someone can understand 'phlogiston' and associate it with T without believing that it is generally associated with T, in the belief that 'phlogiston' is and was generally associated not with T but with somewhat different versions of phlogiston theory. This is clear if T is a strong version of the theory. Even if T is a weak version, they may believe that the word is generally associated with a stronger version, and deny that it is *ipso facto* associated with T. On such grounds, they may even disbelieve that they themselves associate the word with T. Let such sociolinguistic beliefs be false; nevertheless, holding them is quite consistent with understanding 'phlogiston'. It is futile to multiply disjuncts and restrictive clauses in the hope of formulating a sociolinguistic claim so anodyne that anyone who understands 'phlogiston' *must* accept it. The result will just be a complex theoretical claim that ordinary speakers can legitimately doubt, on the grounds that such matters are hard to determine.

A more minimalist line of argument for metalinguistic analyticities appeals to the connection between understanding and knowledge of reference. Suppose that someone understands this sentence:

(1) 'Tree' applies to all and only trees.

Then they understand its constituent words, in particular 'tree' as used at the end of (1). So they know what 'tree' means. For common nouns, knowledge of meaning requires knowledge of application conditions. Consequently, they know that 'tree' applies to all and only trees. Moreover, since knowledge entails belief, they also believe that 'tree' applies to all and only trees. Since they understand (1), they know that it means that 'tree' applies to all and only trees. Thus, it seems, they should knowledgeably assent to (1). The argument generalises to a large class of disquotational claims (disquotationality is crucial, since understanding of the expression used on the right-hand side would not imply knowledge of a truth about a distinct expression mentioned on the left-hand side).

Nevertheless, those who understand (1) may refuse assent to it. Our second imagined speaker is an example, since on his view a universally quantified biconditional with borderline cases for both sides is not definitely true. Indeed, some supervaluationists about vagueness even deny such disquotational principles for vague terms. However erroneous such theories of vagueness, holding them is consistent with ordinary linguistic understanding of (1). If understanding really does involve tacit propositional knowledge of meaning, that knowledge may contradict conscious beliefs.

Let us grant for the sake of argument that understanding (1) entails knowing both that 'tree' applies to all and only trees and that (1) means that 'tree' applies to all and only trees. How then can one understand (1) without assenting to it? We lack direct conscious access to whatever tacit knowledge linguistic understanding is supposed to consist in, otherwise semantics as a branch of empirical linguistics would be much easier than it actually is. We consciously entertain the proposition that 'tree' applies to all and only trees as presented by sentence (1). In tacitly knowing that 'tree' applies to all and only trees (if we do), we tacitly entertain that proposition under a quite different mode of presentation. Thus understanding-assent links such as UA* fail for sentences of natural language:

UA* Necessarily, whoever understands the sentence "Tree' applies to all and only trees' assents to it.

For any tacit assent to the proposition that 'tree' applies to all and only trees is not to it under the modes of presentation that UA* requires.[13]

If linguistic understanding involves tacit propositional knowledge of meaning, it presumably involves tacit assent to the relevant propositions under modes of presentation of some sort. To determine in exactly what sense of 'tacit', if any, understanding does involve tacit propositional knowledge of meaning lies beyond the scope of this chapter. Whatever the answer, no reflective intellectual discipline operates at the level of tacit assent and dissent, even if such a tacit level is necessary for its operation. For instance, linguists' tacit knowledge of their native language does not satisfy the goal of linguistics.

One remaining concern is that logical skills must play some role in linguistic competence because logical features play a role in determining well-formedness. An example is the category of negative polarity items. Consider these sentences:

(2) If she ate any of the cake, she was hungry.
(3)* If she was hungry, she ate any of the cake.

'Any' is a negative polarity item. To a first approximation, the reason why 'she ate any of the cake' is acceptable as the antecedent of the conditional but not as the consequent is that the antecedent is in a downward entailing (negative) context while the consequent is instead in an upward entailing (positive) context. A context C is upward entailing just in case whenever A entails B, C(A) entails C(B); C is downward entailing just in case whenever A entails B, C(B) entails C(A). Thus recognition of the logical features of contexts seems to be needed in order to distinguish between well-formed and ill-formed sentences. But things are not so simple. Consider these sentences:

(4) Exactly four people in the room were of any help.
(5) Few people in the room were of any help.

Logically, 'few' creates a downward entailing context; 'exactly four' does not. However, (4) is acceptable provided that in the context it is taken to imply (5), but not generally otherwise. Thus the phenomenon involves a significant pragmatic element: which contexts are suitable for 'any' cannot be determined on purely logico-linguistic grounds. If we disagree with the speaker of (4) about how many people were in the room or what proportion of them could have been expected to help, we may find her use of 'any' inappropriate without regarding her as *linguistically* incompetent. Similarly, if a speaker has deviant views as to which contexts are downward entailing, but uses 'any' in just those contexts which she regards as downward entailing, we can find her deviant use of 'any' inappropriate without regarding her as linguistically incompetent, precisely because the deviation is explained by logical rather than linguistic unorthodoxy. Thus the role of logical knowledge in such cases does not make it part of purely linguistic competence. All our knowledge is potentially relevant to judging the appropriateness of a given use of 'any'.[14]

4

Understanding words in a natural language has much to do with the ability to use them in ways that facilitate smooth and fruitful interaction with other members of the community. That ability can be realised in indefinitely various forms. Speakers can compensate for their deviance on one point by their orthodoxy on others, their ability to predict the reactions of nondeviant speakers, their willingness in the long run to have their utterances evaluated by public standards. As we have seen, such compensation is often possible when the deviance results from localised interference in the normal practice of using a word by high-level theoretical concerns. Thus there is no litmus test for understanding. Whatever local test is proposed, someone could fail it and still do well enough elsewhere with the word to count as understanding it.

Could an inferentialist reply that such objections trade on a loose everyday sense of 'understanding' that must be replaced by something more precise for theoretical purposes? It is far from clear that a stricter sense would do a better job. The relevant features of the ordinary conception of understanding are not mere unreflective sloppiness. Rather, they are an appropriate response to an important constraint on a theory of linguistic meanings: that there is little point in talking about them unless they can be shared across significant differences in belief, between different individuals at the same time or the same individual at different times. They can survive factual learning and factual disagreement. Although inferentialist accounts respect the letter of that constraint, they violate its underlying spirit, by setting inflexible limits to the scope for genuine disagreement. The more holistic ordinary notion of understanding permits localised disagreement at virtually any point.

Cases of logical deviance hint at ways in which the failure of individualist accounts of meaning go deeper than the immediate lessons of the original anti-individualist arguments of Putnam (1975) and Burge (1979). Their cases are often analysed in terms of a distinction between experts with full understanding and lay-people with partial understanding who defer to the experts, in virtue of which one may correctly ascribe to them attitudes to the contents that experts determine.[15] Such asymmetries are postulated by Putnam's Hypothesis of the Universality of the Division of Linguistic Labour:

> Every linguistic community [...] possesses at least some terms whose associated 'criteria' are known only to a subset of the speakers who acquire the terms, and whose use by the other speakers depends upon a structured cooperation between them and the speakers in the relevant subsets. (Putnam, 1975, p. 228)

But experts themselves can make deviant applications of words as a result of theoretical errors and still count as fully understanding their words. Although they defer to nobody on the matters at issue, they are more than adequately integrated members of the speech community with respect to those very words. Their assignments of meaning to those words are not parasitic on the assignments that more privileged individuals make. Rather, each individual uses words as words of a public language; their meanings are determined not individually but socially, through the spectrum of linguistic activity across the community as a whole. The social determination of meaning requires nothing like an exact match in use between different individuals; it requires only enough connection in use between them to form a social practice. Full participation in that practice constitutes full understanding. That is why there is no litmus test for understanding.

A complex web of interactions and dependences can hold a linguistic or conceptual practice together even in the absence of a common creed that all participants at all times are required to endorse. This more tolerant form of unity arguably serves our purposes better than would the use of platitudes as entrance examinations for linguistic practices.

Evidently, much of the practical value of a language consists in its capacity to facilitate communication between agents in epistemically asymmetric positions, when the speaker or writer knows about things about which the hearer or reader is ignorant, perhaps mistaken. Although disagreement is naturally easier to negotiate and usually more fruitful against a background of extensive agreement, it does not follow that any particular agreement is needed for disagreement to be expressed in given words. A practical constraint on useful communication should not be confused with a necessary condition for literal understanding. Moreover, the practical constraint is holistic; agreement on any given point can be traded for agreement on others. The same applies to principles of charity as putatively constitutive conditions on correct interpretation: imputed disagreement on any given point can be compensated for by imputed agreement on others.[16]

It is far easier and more rewarding to discuss the existence of true contradictions with a dialetheist such as Graham Priest than creationism with a Christian fundamentalist or Holocaust denial with a neo-Nazi.[17] The difficulty of engaging in fruitful debate with fundamentalists or neo-Nazis is not plausibly attributed to some failure of linguistic understanding on their part (or ours); it arises from their wilful disrespect for the evidence. Such difficulty as there is in engaging in fruitful debate with dialetheists provides no significant reason to attribute to them (or us) a failure of linguistic understanding. Competence with the English language no more requires acceptance of some law of noncontradiction or any other logical law than it requires acceptance of the theory of evolution or the historical reality of the Holocaust.

We cannot anticipate all our disagreements in advance. What strike us today as the best candidates for analytic or conceptual truth, some innovative thinker may call into question tomorrow for intelligible reasons. Even when we hold fast to our original belief, we can usually find ways of engaging rationally with the doubter. If a language imposes conditions of understanding that exclude such a doubt in advance, as it were in ignorance of its grounds, it needlessly limits its speakers' capacity to articulate and benefit from critical reflection on their ways of thinking. Such conditions are dysfunctional, and natural languages do not impose them.[18]

There is, of course, a distinction between understanding a word and not understanding it. One can lack understanding of a word through lack of causal interaction with the social practice of using that word, or through interaction too superficial to permit sufficiently fluent engagement in the practice. But sufficiently fluent engagement in the practice can take many forms, which have no single core of agreement.[19]

5

At this point, someone sympathetic to the spirit of epistemological analyticity may suspect that the mistake was to go for the idea that understanding is somehow *psychologically* sufficient for assent. Instead, the suggestion is, we should go for the idea that understanding is somehow *epistemologically* sufficient for assent.[20] Externally, logically deviant speakers are in a position to know (or to assent with justification). They seem to be wilfully and perversely turning their backs on knowledge that is available to them. It is there for the taking, but they are psychologically blocked from taking it.

We must be careful about the source of the blockage. Suppose that it is lack of logical insight. Although our imagined speaker understands the word 'and', he lacks the logical insight competently to apply conjunction elimination. Other people just like him except for having more logical insight do competently apply conjunction elimination. Anyone who understands 'and' and has a modicum of logical insight can competently apply conjunction elimination. That story assigns no special role to linguistic understanding, beyond the usual role that it plays as a precondition for all intelligent discourse: the decisive role is assigned to logical competence, not semantic competence.

For semantic competence to play the decisive role, something like this is needed:

KU Whoever competently applies conjunction elimination for 'and' in the normal way does so simply on the basis of their understanding of 'and'.

(Treat 'on the basis of' more like 'by an exercise of' than like 'by inference from'.) KU may be plausible at first sight. It does not imply that whoever understands 'and' applies conjunction elimination for 'and' or even has a disposition to do so, competently or otherwise.

What does the definite description 'their understanding of "and"' in KU denote? There are thick and thin candidates. The thin candidate is the mere fact that they understand 'and'. The thick candidate is the body of underlying facts that constitute the thin candidate, the facts that realise this particular subject's understanding of 'and' at this particular time. The thin candidate is exactly similar for any two people who understand 'and', since they have the same property of understanding it. The thick candidate may differ between any two people who understand the word, since different underlying facts can constitute their doing so. These characterisations are schematic, but will do for present purposes.

Suppose that the definite description in KU denotes the thick candidate. KU remains plausible on this reading. Then, given the picture of linguistic understanding in previous sections, KU has much less epistemological significance than might have been hoped. The facts that constitute your understanding of a given word include various cognitive capacities that are not in general necessary for understanding that word, but help to make up your particular competence with it. For example, the facts that constitute the understanding of 'and' on the part of the imagined speaker who rejects conjunction elimination for 'and' include his logical capacities and dispositions. The basis cited in KU includes cognitive capacities that are not in general necessary for understanding the word. Thus the thick candidates are too thick to yield bases for analyticity; they involve knowledge that is not semantic in any relevant sense.

Suppose instead that the definite description in KU denotes the thin candidate. But it is not the basis in any useful sense for competently applying conjunction elimination for 'and' in the normal way, although confusion with the thick candidate may suggest otherwise. The thin candidate implies no specific logical capacity at all. It is not as though in such cases a subject's understanding quietly tells them to make the inference but they override the advice; it is providing no such advice to be overridden. Consequently, understanding in the thin sense provides no basis for assent to anyone. Of course, understanding 'and' is a precondition for competently applying conjunction elimination for 'and', and in that sense may be *part* of the basis for so applying it, but that point is quite general; it is neutral between the analytic and the synthetic. Thus the thin candidate is too thin to be a basis for knowledge.

We could try eliminating the talk of bases, for instance in a formulation like this:

AJ Whoever understands 'and' and applies conjunction elimination for 'and' does so with justification.

But such principles are false, since someone who understands 'and' but applies conjunction elimination for 'and' because his manifestly unreliable guru told him to does so without justification. The obvious way to avoid such counterexamples and make the connection with semantic competence is to qualify 'applies conjunction elimination' by 'on the basis of that understanding'. But that returns us to the problems of KU.

The problem is general. The idea that, in the cases at issue, understanding is epistemologically sufficient for assent to the inference is the idea that assent on the basis of understanding has the desired positive epistemic status. But once we disambiguate 'understanding' between thick and thin candidates, we can see that the thin candidates are too thin to be bases for assent while the thick candidates are not purely semantic or conceptual. The attempt to base the epistemology of obviously valid inference rules such as conjunction elimination on preconditions for linguistic competence with them rests on a confusion between linguistic competence and logical competence. Similarly, dissent from the law of noncontradiction in its various forms is simply a logical failure; dialetheists such as Graham Priest manifest semantic competence with the logical particles of their native languages in their ordinary senses. Epistemological analyticity is an illusion.

Notes

* I thank Diego Marconi for a quarter-century of philosophical conversations and friendship. A fuller version of the arguments in this chapter can be found in (Williamson, 2008), which in particular assesses the present standing of Quine's critique of the analytic-synthetic distinction and develops more fully a positive account on which shared meanings do not require agreement on anything in particular. The acknowledgements there are relevant to this chapter, but I should particularly mention the discussion of related material at the 2006 conference at Reggio Emilia on the question 'Is there anything wrong with Wittgenstein?,' co-organised by Diego, at which I was present as a specimen anti-Wittgensteinian, the equivalent of the psychopath whom psychoanalysis cannot reach (I have received Wittgensteinian therapy on average at least once a month since 1973 without showing any sign of improvement). Thanks to Alberto Voltolini for helpful comments on an earlier version of this chapter.

1. The case of deductive logic is a useful reminder that many short, trivial steps of no apparent significance can be chained together into a long, nontrivial argument of obvious significance: the short steps were not really insignificant after all: no apologies for concentrating on them here.
2. In effect, Horwich (1998b, pp. 131–53) allows understanding-assent links for which the understanding-truth links fail.
3. See Eklund (2002) for a defence of the idea of inconsistent languages.

4. Dummett (1973, pp. 397, 454) claims that the rules for pejorative terms such as 'Boche' suffer from a related kind of incoherence; Brandom (1994, p. 126; 2000, pp. 69–70) and Boghossian (2003, pp. 241–2), amongst others, have relied on his description of the practice of using such terms. I argue that it is mistaken in Williamson (2003, 2009), and suggest an alternative.
5. See Peacocke (1992, p. 21) and Boghossian (2002a).
6. Cf. Boghossian (2003a, pp. 242–3), representing a change of view from Boghossian (2002a).
7. The treatment of the issue in Boghossian (2003a) is of this general kind. For detailed criticism see Williamson (2003).
8. In discussion, Boghossian suggested conjunction elimination as a fall-back example of a nondiscretionary rule if modus ponens fails. Peacocke writes of the possession-condition for the concept of conjunction, 'On any theory, this possession-condition will entail that thinkers must find the transition from A and B to A compelling, and must do so without relying on any background information' (Peacocke, 2004, p. 172).
9. This conclusion contrasts with the discussion of paraconsistent logic in Marconi (1987, pp. 73–96). Williamson (1989) responds critically to that discussion.
10. The seminal paper is Tversky and Kahnemen (1983). We can also imagine speakers who reject instances of conjunction elimination through muddling truth and conversational appropriateness. 'Did she take the money and give it back? Yes. Did she take the money? No, she took-the-money-and-gave-it-back.'
11. Compare Harman (1999, p. 151) on problems in analysing 'bachelor' as 'unmarried adult male' and Nozick (2001, pp. 135–6) on the nonsynonymy of 'mother' and 'female parent'.
12. One problem with interpreting speakers as all speaking their own idiolects is that it tends to undermine testimonial knowledge: if Y gets some knowledge from X and passes it on to Z in the same words, they do not mean in Y's mouth what they meant in X's.
13. See also Soames (1995) for relevant considerations.
14. Ladusaw (1996, pp. 325–7) surveys issues concerning negative polarity.
15. An example is Peacocke's (1992, pp. 29–33) discussion of deference-dependent propositional attitude ascriptions. Burge (1986) extends his earlier arguments in ways related to the arguments of this paper, in his account of the understanding of words such as 'sofa', and argues for such a deeper lesson. Goldberg (2000) replies on behalf of Burge to Bach (1988) and Elugardo (1993). Although the discussion of Putnam and Burge's semantic anti-individualism in Marconi (1997) is hostile, Williamson (1998) argues that much in Marconi's account of lexical competence can be reconciled with semantic anti-individualism once, following Burge, it is combined with psychological anti-individualism, as it needs to be.
16. Davidson (2001, pp. 144–9) famously endorses a holistic principle of charity while rejecting the analytic-synthetic distinction. See Williamson (2008) for more discussion of charity. Of course, he takes the notion of a shared language less seriously than here (Davidson, 1986).
17. For examples of rational debate for and against a law of noncontradiction see Priest, Beall and Armour-Garb (2004).

18. W. B. Gallie's intriguing account of the positive function of 'essentially contested concepts' is relevant here; his examples are 'the concepts of a religion, of art, of science, of democracy and of social justice' (Gallie, 1964, p. 168).
19. Someone who understands a word without being disposed to utter it (perhaps because they find it obscene or unpronounceable) can still count as sufficiently engaged in the practice of using it. The account should also be read so as to allow for understanding of dead languages.
20. Some rationalist defenders of intuition seem to have something like this in mind.

Part III

Metaphilosophy

10

An Adequacy Condition for the Interpretation of the *Tractatus* Ontology

Pasquale Frascolla

For all those who don't take the *Tractatus* as an extraordinarily ingenious and extremely ironic exercise in Kierkegaardian style, the problem of the interpretation of its ontology still stands as one of the most prominent among the unsolved, or perhaps one should say 'unsolvable,' questions surrounding that text. As known, it is not hard to extract from the *Tractatus* a skeletal ontological schema, in which a general characterization of the entities playing the semantic role of references of names (objects) and of the entities playing the semantic role of senses of elementary propositions (states of affairs) can be given. No serious interpreter of the *Tractatus*, however, should be content with that ontological skeleton: too many explicit statements which are to be found in Wittgenstein's early masterpiece are doomed to remain unexplained and desperately obscure if the interpreter accepts to confine him or herself to the limits of a schematic ontology. On the other hand, a strong impulse to the acceptance of those limits originates from other equally well known aspects of Wittgenstein's position: first, the mere but insurmountable fact that no instance either of an object or of a state of affairs is given in the *Tractatus*; second, Wittgenstein's clear and definite statement that it is the application of logic, not pure logic alone, that is charged with the task of discovering the possible forms of elementary propositions; last but not least, his almost resentful answer to a question put by Malcolm many years later, that 'it was not his business, as a logician, to try to decide whether this thing or that was a simple thing or a complex thing, that being a purely empirical matter.'[1]

In my opinion, no one of those three aspects entails that one should give up the attempt at fleshing out the ontological schema, at putting flesh on the bones, so to speak, thus transforming the colourless skeleton into a richer ontology, which to my mind is to be understood as a full-blown phenomenalistic ontology.[2] Here I will take for granted that that transformation can be done and I will focus on a different matter: is it possible to identify, in that portion of the text of the *Tractatus* specially devoted to ontology, a group of theses which, taken together and consistently accounted for, actually enable the interpreter to go well beyond the skeletal ontological schema? My definite answer to the above question is, yes, it is. The strategy I intend to adopt in order to justify this positive answer consists in setting up a requirement of material adequacy for the interpretation of the *Tractatus* ontology, that is, a condition that any tentative interpretation which aims at being assessed as a sound interpretation ought to fulfil. To say that an interpretation is sound does not mean, of course, that it is *the* correct interpretation: that would be too much. Nonetheless, any interpretation which proves not to be sound because of its incapacity to meet the requirement of material adequacy should safely be discarded as incorrect.

The adequacy condition runs as follows. From any sound interpretation of the *Tractatus* ontology, each one of the five theses (a) – (e) below should be deducible:

(a) the status of substantiality accorded to objects is compatible with the possibility that an object does not occur in anyone of the facts into which the world divides (TLP 2.013);
(b) objects are colourless and, by a natural generalization, they are not-spatial and timeless as well (TLP 2.0232);
(c) space and time are on a par with colour as forms of objects (TLP 2.0251);
(d) the Principle of Identity of Indiscernibles does not hold of objects (TLP 2.0233, 2.02331, 5.5302);
(e) no relation proper connects objects in a state of affairs (TLP 2.03–2.032).

In framing theses (a) – (e), I have put in brackets the reference to the sections of the *Tractatus* which are relevant for them. The double task that I want to accomplish here is: to show how those sections, along with some other crucial ones, provide strong evidence for the ascription to Wittgenstein of the theses in question; at the same time, to make clear the content of those theses within the framework of the *Tractatus*, and to explore their ramified implications and their far-reaching consequences as to the question of the nature of objects, states of affairs and facts.

Let us start from thesis (a), which constitutes the true architrave of the entire edifice. The first step towards the clarification of its content consists in unpacking the metaphor of logical space. In fact, in section 2.013 Wittgenstein speaks of the 'space of possible states of affairs' in which each thing (object) is placed and says that 'this space I can imagine empty,' although 'I cannot imagine the thing without the space.' One can reasonably assume that logical space is the totality of logical places, but soon the further inevitable question arises, what is a logical place? According to Wittgenstein, every meaningful proposition determines a logical place (TLP 3.4), and this statement can be construed as saying that it is the state of affairs depicted by a meaningful (elementary) proposition that is metaphorically described as a logical place. Just as physical space is constituted by physical places, so logical space is constituted by logical places, the latter being states of affairs, possible combinations of objects. The metaphor can smoothly be extended by comparing the obtaining of a state of affairs with the being filled up by matter of a physical place, and this extension throws a vivid light on its true import: the obtaining of a state of affairs can metaphorically be described as the being filled up of a logical place, in the sense that, whenever a state of affairs obtains, one can speak of a fact that fills up a logical place. Out of metaphor, this means that a state of affairs embodies the very possibility of a corresponding fact: 'In geometry and logic alike a place is a possibility: something can exist in it' (TLP 3.411). By adopting the metaphor, one is enabled to say that a certain logical place is filled up by matter, instead of saying, literally, that a certain state of affairs obtains, is a fact, and that a certain logical place is empty, instead of saying, literally, that a certain state of affairs doesn't obtain, isn't a fact.

The metaphor nicely works if one aims at characterizing all the conceivable configurations of reality, all possible worlds, in the only sense in which Leibniz's notion can be accommodated within the framework of the *Tractatus*. Given the whole logical space, which is as fixed and unalterable as objects are, reality is nothing but the actual distribution of matter over logical places, a distribution in which some of them are filled up by matter whereas the remaining ones are left empty, that is, out of metaphor, reality is the actual configuration of the obtaining and non-obtaining of states of affairs. Any other conceivable configuration of reality differs from the actual one either because some (at least one) among the logical places which are actually filled up by matter are empty in it, or because some (at least one) among the logical places which are actually empty are filled up by matter in it, or for both. Literally, this means that any conceivable configuration of reality differs from the actual one either because some (at least one) among the actually obtaining states of affairs do not obtain in it, or because some (at least one) among the actually non-obtaining state of affairs do obtain in it, or for both.

Equipped with the understanding of the metaphor of space, let us turn to section 2.013. What is the space in which each object is placed? The answer is straightforward: we have just shown that to speak of a logical space amounts to speaking of a set of logical places and hence, unpacking the metaphor, to speaking of a set of states of affairs. Wittgenstein maintains that each object is to be conceived of as placed in the space constituted by all those states of affairs of which it is a constituent: since an object cannot be thought of as an isolated entity, but only in concatenation with other objects, that is, in states of affairs of which it is a constituent, and since all its possibilities of combination with other objects in states of affairs are given with the object itself, are written into it, no object can be imagined without its space. But what does it mean that one can imagine empty the space where an object is placed? Here too the answer is straightforward: we have just seen that an empty logical place is a logical place which is not filled up by matter, that is, literally, a non-obtaining state of affairs; by naturally extending the metaphor, a space is empty if all the logical places which form it are empty. Therefore, to imagine empty the space where an object is placed is tantamount to imagining that no one of the states of affairs which are the places constituting that space obtains. In conclusion, Wittgenstein is making the claim that nothing can prevent us from imagining that no one, among the states of affairs of which an arbitrary given object **o** is a constituent, obtains, or, in other words, that nothing can prevent us from imagining that **o** does not occur in anyone of the facts into which the world divides. Leaving imagination apart, the point can be put in clearer terms as follows: among the possible worlds, that is, among the possible configurations of the obtaining and non-obtaining of states of affairs, there are many in which no state of affairs containing the object **o** as one of its constituents obtains, for every object **o** (all those combinations which leave empty the space where **o** is placed).

The theoretical relevance of the preceding conclusion can be appreciated only if one realizes that, at least apparently, it clashes with the well known thesis that 'objects make up the substance of the world' (TLP 2.021). Objects are what remains constant in logical variation, are so extraordinary things that their existence is not affected by any logically possible change in the configuration of the world, by any passage from one conceivable configuration of the obtaining and non-obtaining of states of affairs to any other one: this is what their admirable property of substantiality amounts to. But how can that status be reconciled with the acknowledgement that, for any arbitrary object **o**, there are some configurations of reality in which, so to speak, **o** disappears, in the sense that no fact in that configuration, no state of affairs which obtains in that configuration, contains that object as a constituent? (don't forget that objects cannot occur in the world as isolated entities). What stuff are objects to be made of, if they must combine two at least apparently quite contradictory properties, as that of being unaffected by logical variation and that of possibly not occurring in any fact? It seems to me that, for a full understanding of the *Tractatus* ontology, these questions call for an answer and cannot be dismissed as futile: *hic Rhodus, hic salta*.

I will not try to give a direct answer to the above questions, but I will follow a more indirect path by focusing on theses (b) and (c) of the preceding list, before coming back to the problem which has been provisionally left open. In section 2.0231, Wittgenstein says that material properties of the world are produced only by the configuration of objects and are represented only by propositions. Since elementary propositions depict states of affairs in which objects are combined together, we can safely say that it is the obtaining of a state of affairs or, what is the same, the existence of a complex that brings about the fact that the world enjoys a corresponding material property. For instance, the being red of a certain area of the (physical or visual, it doesn't matter) space is a material property of the (physical or phenomenal) world and is brought about by the realization of a certain combination of objects, by the existence of a certain complex. On the other hand, objects, insofar as they lack internal constituents, cannot be red, or, by generalizing the point to all colours, cannot be coloured. A very important double conclusion as to the nature of objects and states of affairs (complexes) can be drawn from the preceding considerations: at least some entities among those classified as objects, by combining with other objects produce coloured complex entities, whose contingent existence is affirmed by elementary propositions; at the same time, all objects are colourless. Colours, of course, can produce coloured entities without being themselves coloured, and hence are good candidates for the role of objects.

A point perfectly analogous to that made for colours can be made, in my opinion, both for spatial places and moments of times. Notice that in what follows no distinction is presupposed neither between physical and visual space, nor between physical and phenomenal time. A given entity's being located at a certain place (at a certain moment of time) is a material property of the (physical or phenomenal) world and it is to be traced back to the realization of a determinate contingent combination of objects, to the contingent existence of a complex. The same holds of temporal position: a given entity's being positioned at a certain moment of time (at a certain place of space) is a material property of the (physical or phenomenal) world, and again it is the realization of a determinate contingent configuration of objects, the contingent existence of a complex, that produces that property. On the other hand, objects, being simple, cannot have either a location in space or a position in time. The same double conclusion which has been drawn earlier with reference to colours can be drawn now with reference to spatial places and moments of time: at least some entities among those classified as objects, by combining with other objects produce spatial complex entities, but no object occupies a spatial place; moreover, at least some entities among those classified as objects, by combining with other objects produce temporal complex entities, but all objects are timeless. Spatial places and moments of time, of course, can produce space-temporally located complex entities without being themselves space-temporally located, and hence they are on a par with colours (colourless entities which, by combining with other objects, produce coloured complexes) as good candidates for the role of objects.

Before crossing the threshold of the analysis of thesis (c) of our list, let us pause for a moment to better assess the import of the preceding conclusions. From a metaphysical point of view, complexes, whose existence is affirmed by elementary propositions, are particulars, because of their location in space and/or their position in time which makes them unrepeatable entities.[3] But the application of the atomistic model gives a peculiar twist to the way in which their particularity is realized: we saw that to say that a certain complex is red is tantamount to saying that red is one of its constituents; for the same reason, to say that it occupies the place *l* means that that place is another one of its constituents, and to say that its position in time is *t* means that that moment is still another one. Though in disguise, the problem we are facing is the old problem of individuation. To clarify matters, it is expedient to imagine the following situation: let us suppose that we have to do with phenomenal complexes, such as visual coloured spots; more specifically, let us suppose that two red round spots occupy at the same time two different places of the visual field. If phenomenal qualities such as those of being red and of being round were the only simples, that is, the only constituents of the two complexes, then the two complexes would have the same constituents and hence they would be numerically one and the same complex, which would be absurd. An alternative between two radically different solutions to the problem of individuation arises here: according to the first one, those momentary entities which are bare particulars (temporal cross sections of visual areas, for example) are given an ontological status, and the distinction between the two spots is thus ensured by the purely numerical difference of the bare particulars which are included in the two spots and which are the different bearers of the properties of being red and of being round; according to the second solution, the phenomenal qualities of being at a certain place of the visual field and of being at a certain moment of phenomenal time are introduced into ontology and are dealt with as further qualitative constituents of complexes. Accordingly, the two red round spots would be complex particulars, whose individuality is secured by the difference in their spatial qualitative constituents.

An influential exegetical tradition has maintained that Wittgenstein in his early masterpiece chose the first alternative: as a consequence of that choice, states of affairs would be conceived of as composed of bare particulars and of properties or relations, that is, universals instantiated by the former. In my opinion, section 2. 0251 settles the matter in favour of the second alternative, the one according to which particulars are to be conceived of as combinations of qualities, including spatial locations and temporal positions.[4] The key-argument supporting my claim rests on the interpretation of the notion of form of an object. The form of an object can be characterized as its combinatorial potential, its capability to combine with other objects in those broader structures which states of affairs, or complexes, are. One can extensionally identify the form of an object **o** with the set of all the n-1-uples (for n ≥ 2) of objects with which it can combine in states of affairs; accordingly, two objects have the same form if they can combine with the same n-1-uples (for n ≥ 2) of other objects, and differ in form otherwise. When the objects $\mathbf{o}_1, \mathbf{o}_2,\ldots, \mathbf{o}_n$ have the same form **F**, in the sense just explained, one can say, alternatively, that the form **F** is a category of objects which collects together the objects $\mathbf{o}_1, \mathbf{o}_2,\ldots, \mathbf{o}_n$ enjoying the same combinatorial possibilities. In counting *colour* among the forms of objects, Wittgenstein is simply stating that red, blue, yellow etc. are objects which share the same combinatorial possibilities and which, for that very reason, are to be classified under the same formal category, that of *colour.*

What is really decisive for our purposes is the fact that even *space* and *time* are both numbered among the forms of objects and thus are dealt with on a par with the form *colour.* By arguing in the same way as before, we are allowed to affirm that places $\mathbf{l}_1, \mathbf{l}_2, \ldots, \mathbf{l}_n$ are objects having the same form because each one of them can combine with the same n-1-uples (for n ≥ 2) of other objects, and that the form *space* is nothing but the category of objects that collects them all together. Similarly, we can conclude that moments $\mathbf{t}_1, \mathbf{t}_2, \ldots, \mathbf{t}_n$ are objects having the same form because each one of them can combine with the same n-1-uples (for n ≥ 2) of other objects, and that the form *time* is nothing but the category of objects that collects them all together. A particular is generated by combining together one object from each one of a certain number of different categories (forms) of objects: for instance, a coloured complex is generated by combining all together one colour, one place and one moment of time.

Many important consequences follow from the above reading of section 2.0251, and not only on the ontological side. First of all, the *Tractatus* complexes or states of affairs are not to be construed in conformity with the traditional dualistic model of a substance which is not reducible to the sum of its attributes, and which, at the same time, makes individuation possible, that is, a bare particular, on the one hand, and of attributes which inhere in that substance, on the other. Complexes result from the combination of qualities, and individuation is secured by the presence of places and/or moments of time among those qualities: certainly not by chance, it is precisely the conception of complexes as bundles of qualities that Wittgenstein sharply criticizes in some of the writings of his so-called intermediate phase, overtly devoted to a discussion of his early ontological views.[5] The counterpart, on the logico-linguistic side, of the rejection of the substance/attribute schema ought to be the rejection of any privileged role of the subject/predicate schema in the determination of the form of elementary propositions. But more on this theme will come later, in discussing thesis (e) of our list.

For now, let us focus on another relevant aspect of the conception of particulars, or complexes, that I am ascribing to Wittgenstein. According to the interpretation based on the substance/attribute model, in a state of affairs one or more particulars exemplify one universal (property or relation). Once that the conception of complexes as combinations of qualities is adopted, a completely new view of exemplification emerges: let us suppose that the colour $\mathbf{c}_1$, the place $\mathbf{l}_1$ and the moment $\mathbf{t}_1$ are the constituents of a complex **C**; then, one can say that **C** has the qualities $\mathbf{c}_1$, $\mathbf{l}_1$ and $\mathbf{t}_1$, or that **C** is an instance of the three qualities. In more formal terms, the relation of *being an instance of* can be defined as the converse relation of the relation of *being a constituent of*: for example, a red complex is an instance of redness because redness is one of its constituents, and a complex located at $\mathbf{l}_1$ is an instance of the place $\mathbf{l}_1$ because $\mathbf{l}_1$ is one of its constituents. Complexes, which are particulars, unrepeatable entities, exemplify all the qualities that are their constituents. Qualities, on the contrary, are repeatable entities, because, for example, space-temporally different complexes can have the same colour quality as a constituent, just as differently coloured complexes can have the same place-quality as a constituent (and, in that case, different time-qualities, due to the incompatibility of colours): plainly, qualities are universals, and thus objects, which are identified with qualities, are themselves universals.

At last, we have reached a position that enables us to find out the solution to the problem we came across at the end of our comment upon thesis (a) of our list, that is, the problem of how objects can possess two at least apparently quite contradictory properties, as that of being unaffected by logical variation and that of possibly not occurring in any fact. It is their very nature as universals that enables objects to be simultaneously endowed with those two properties: in the framework of the *Tractatus* atomism of qualities, to say that an object/quality does not occur in any fact simply means that there are no existing complexes, no obtaining states of affairs, which are instances of that object/quality. Such a circumstance does not jeopardize the 'existence' of the object/quality, in the sense that it is a constituent of states of affairs which could be instanced by one or more facts in different configurations of the obtaining and non-obtaining of the totality of states of affairs. There is a further related point which deserves close attention, nonetheless. It deals with the semantic role of objects as references of names: since names are not susceptible of being verbally defined, and acquire their meaning by direct association with their reference, one is compelled to suppose that every object has been instanced by some existing complex at least once or, in other words, that, for every object **o**, at least one state of affairs of which **o** is a constituent has been obtaining, has been a fact. It is worth noting that in Wittgenstein's later critical discussions on the semantic role of ostensive definitions, the main issues, though framed in a very different jargon, derive from the circumstance that entities referred to in those procedures (e.g., colours) always occur as constituents of existing complexes, not in isolation.

If both the nature of objects as universals and their role as constituents of complexes which exemplify them are recognized, two immediate corollaries concerning the problem of the validity of Leibniz's Principle of the Identity of Indiscernibles in the *Tractatus* can be derived and, with them, thesis (d) of our list can be justified. A clear-cut distinction is to be made between the case of simple entities (objects/qualities) and the case of complexes. It is easy to prove that Leibniz's Principle analytically holds of complexes, or states of affairs. We know that, for a complex **C**, having the quality **Q** means that **Q** is a constituent of **C**; now, two indiscernible complexes $\mathbf{C}_1$ and $\mathbf{C}_2$ share all their qualities and thus they have the same constituents; but two complexes with the same constituents are one and the same complex, given the purely extensional nature of complexes. Things radically change if we turn to objects. It is expedient here to start from Wittgenstein's distinction between formal (or internal) and material (or external) properties of objects. To say that an object **o** has the material property **P** means that **o** is a constituent of a determinate *obtaining* state of affairs, or of a determinate *existing* complex. For example, if a red complex exists at the place **l** at the moment **t**, one can say that the property of being combined with the place **l** and with the moment **t** is a material property of the object redness. Notice that this does not mean that the object redness is at the place **l** at the moment **t**: being combined with **l** does not mean being at **l**, and being combined with **t** does not mean being at **t**. Redness is an abstract universal and has no location either in space or in time, whereas what is located at **l** at **t** is the complex which is an instance of redness, of **l** and of **t** by having them as its constituents. Formal properties of an object, on the other hand, are to be traced back to its possibilities of combination with other objects in states of affairs, or complexes: for example, it is one of the formal properties of the object redness that it *can* combine with the place **l** and with the moment of time **t** to generate a red complex.

Plainly, indiscernible objects are objects that share both all their formal properties and all their material properties. Wittgenstein's rejection of the applicability of the Principle of the Identity of Indiscernibles to objects amounts to the claim that two numerically different objects could share all their properties, the formal ones and the material ones. That two different objects can have all their formal properties in common follows from the interplay of our characterization of formal properties and of our definition of the form of an object. It is crystal clear that any two objects with the same form, that is, falling under the same formal category, share all their formal properties. For instance, two different colours can combine with exactly the same n-1-uples ($n \geq 2$) of other qualities to form coloured complexes and, in force of their having the same combinatorial potential, they cannot be distinguished by means of their formal properties alone. But nothing prevents us from imagining that two objects with the same formal properties have all their material properties in common as well: two visual places, for example, could be combined with the same visual colour at every moment of phenomenal time.[6] In conclusion, two objects, even though formally and materially indiscernible, would continue to be two, contrary to Leibniz's Principle, on which Russell's definition of the relation of identity was in turn based. Notice that Wittgenstein's solution to the problem of individuation is the key of the argument: abstract universal entities – and objects, insofar as they are qualities, are such – have no space-temporal location and thus one cannot resort to the latter in order for a difference in their properties to be recognized whenever the objects are two.

In approaching the last thesis of our list (thesis (e)), a quick philological detour is expedient. It deals wit the strange destiny of the English word 'relation' which has been used to translate at least three different German expressions occurring in the *Tractatus*, with very dangerous effects, in my opinion. The most striking example is that of the pair of sections 2.03 and 2.031, the latter being, according to Wittgenstein's system of enumeration, the first comment upon the former. In 2.03 the famous metaphor of the links of a chain is introduced to illustrate the way in which objects are to be thought of as combined together in a state of affairs or complex. No doubt, in the first comment upon 2.03, that is, in section 2.031, it is the same idea that is expressed again. The English translation, however, runs as follows: 'in a state of affairs objects stand *in a determinate relation* to one another.'[7] Now, 'stand in a determinate relation to one another' translates the German text '*verhalten sich.... in bestimmter Art und Weise zueinander*,' but the use of the substantive 'relation' in the English translation threatens to reverse the true meaning of the metaphor of the chain, that is, that no relation proper connects objects in a state of affairs. The same word 'relation' has sometimes been used to translate the German word '*Beziehung*,' which occurs in many passages of the *Tractatus* where the relationship between language and reality is dealt with, and, what is more important for our purposes, it has been obviously used to translate the German word '*Relation*' as well, by thus creating the appearance that in states of affairs objects entertain ordinary relations, that is precisely what the metaphor of the chain denies.

The idea of objects as entities entertaining genuine relations in a state of affairs has shaped the profile of a significant part of the exegetical debate. As is well known, many different interpretations have been devised in order to account for the role of relations, and more generally of universals, in the internal constitution of states of affairs.[8] If the view of states of affairs as immediate combinations of qualities is accepted, only one alternative remains: either qualities are combined together in a state of affairs exclusively in virtue of their combinatorial potential, of their form, and then the form itself of the state of affairs comes to be reducible to that of its constituents, without residue; or a sort of pseudo-relation of *being combined together* is introduced as the pervasive glue which would be present in every state of affairs, but which would not play the role of a further constituent of it. The choice between the two lines of thought is not devoid of consequences on the way elementary propositions are conceived. According to the latter line, in an elementary proposition *n* abstract singular terms would be concatenated with an *n*-adic pseudo-relational term; according to the former, no resemblance would survive between the elementary propositions of the *Tractatus* and the atomic propositions of Frege's and Russell's logical calculi. It cannot be passed in silence that in his 1929–33 reflections on the nature of elementary propositions, Wittgenstein repeatedly expresses his conviction that the form of elementary propositions could be very far from the model of atomic propositions of the logicist tradition. In conclusion, a clarification of the whole problematic concerning the relationship between objects and states of affairs or complexes in the *Tractatus* can be achieved only on condition of disentangling the question of the status of the nexus linking objects in a state of affairs from the question of the role of universals in its ontology. The fact that no relation in any ordinary sense connects objects in a state of affair is the premise on which that work of disentanglement is to be based, even though the recognition of that fact, taken alone, does not suffice to settle the matter as to the internal composition of states of affairs (only particulars? particular(s) plus one universal? only universals?).[9]

The meticulous exposition of the content of theses (a) – (e) of our list aims at persuading the reader that no interpretation of the *Tractatus* ontology can be accepted as sound unless it consistently accounts for all of them. Any interpretation able to do that would meet the requirement of material adequacy I set up at the outset of the chapter. As remarked earlier, however, meeting that requirement is not enough, taken alone, to univocally determine 'the' correct interpretation. We spoke of the ontology of the *Tractatus* as an atomism of qualities, but nothing has been said so far on the nature of those qualities: are they physical qualities or phenomenal qualities? The answer to this question comes from an area of the *Tractatus* different from that specially devoted to ontology, that is, the area where solipsism is introduced and the explicit identification of the world with non-physiological and non-psychological life, of the macrocosm with the microcosm, is made. Since many interpreters among those who accept only the skeletal ontological schema are not willing to acknowledge that the world which is spoken of at the outset of the *Tractatus* is the same world which later in the text is said to be 'my world,' it was extremely important to show that, quite independently of solipsism, the *Tractatus* ontology entails theses (a) – (e) of our list and that its full understanding can be achieved only on condition of explaining them all at one stroke.

Once that the world is identified with what is immediately given to a non-empirical subject, with the stream of phenomena, the largely hidden ontology of the *Tractatus* can conjecturally be made explicit. Objects are to be identified with repeatable phenomenal aspects or qualities, *qualia* in Nelson Goodman's sense of the word, whereas states of affairs are to be identified with phenomenal complexes, which belong to the various sense realms and which have *qualia* as their ultimate constituents. As a program for the 'new' philosophy that the *Tractatus* tended to inspire, the application of logic should show, among other things, how the physical world can be constructed out of the phenomenal world by means of logical machinery. It is easy to show that the phenomenalistic ontology based on *qualia* fulfils the adequacy condition framed in the paper; moreover, as far as I know, no other interpretation of the *Tractatus* ontology succeeds in this equally well.[10]

Notes

1. (Malcolm, 1958, p. 70).
2. See Frascolla (2007, Chapter 3).
3. The 'and/or' alternative is justified by my intention of leaving unprejudiced the question concerning the physical or phenomenal nature of complexes: temporal not-spatial complexes can be found in phenomenal world (auditory complexes, for example), not in physical universe.
4. Gustav Bergmann and Erik Stenius, among others, subscribed to the first interpretation of the *Tractatus* ontology (see Bergmann (1964) and Stenius (1960)). I am attributing to Wittgenstein a conception of particulars which is akin to the one put forward by Russell many years later, according to which particulars are bundles of coexisting qualities (see Russell (1940)).
5. See, for example, the well known appendix to *The Big Typescript* entitled 'Complex and Fact.'
6. I am not sure that the same claim could peacefully be made with reference to physical places. Moreover, that two visual colours cannot be combined with the same visual place at every moment of phenomenal time is a consequence of the incompatibility of colour attributions. This limitation,

in Wittgenstein's hopes, could be overcome through an analysis of shades of colour into their internal components.

7. (Wittgenstein, 1922), English translation by D. F. Pears and B. F. McGuinness, italics mine.
8. Any reference to the nominalistic exegetical tradition, which takes the only constituents of states of affairs to be particulars, will be skipped here.
9. On this theme see G. Lando's doctoral thesis.
10. See Frascolla (2007).

11
Method and Metaphilosophy in the *Philosophical Investigations**

Meredith Williams

One of the fundamental debates over how to understand Wittgenstein's *Philosophical Investigations* concerns whether he is developing theories of language and mind that are alternatives to traditional theories. His explicit metaphilosophical remarks state otherwise. Famously Wittgenstein insists that he offers no theories, he provides no explanations of the phenomena philosophers have been preoccupied with; rather 'philosophy may in no way interfere with the actual use of language; it can in the end only describe it' (PI § 124). Yet much in Wittgenstein's later writings can plausibly be taken to offer an alternative account of language and mind. So, on the one hand, we find constructivist interpretations that can only downplay the significance of his metaphilosophical remarks. And, on the other hand, we find quietist interpretations that privilege the metaphilosophical remarks. The constructivist tends to view Wittgenstein's style of philosophizing as something more idiosyncratic to the man and so not essential to pursuing his critical and positive insights. The quietist, on the other hand, takes the style of arguing to be the essential feature of Wittgenstein's task, which is to change in a profound way our entire attitude towards the philosophical project. We are to replace the theoretical drive of traditional philosophy with a piecemeal struggle to overcome the urge to theorize, a struggle that achieves its end when we recognize how misplaced the drive to theory really is. The constructivist reads Wittgenstein as continuous with the philosophical tradition, though offering radically new theories, while the quietist sees a goal that requires a radical break with the tradition.

Yet we need not be forced to choose between these two interpretations. A therapeutic methodology of the sort emphasized by Wittgenstein's metaphilosophical remarks does lie at the heart of his project. The constructivist goes wrong in attributing to Wittgenstein theories that rival those of Carnap or Russell. But the quietist goes wrong in her insistence that Wittgenstein does not offer us a new picture of language and mind. Both interpretations work with a restricted conception of (positive) philosophy, one that sees philosophical theorizing as of a piece with scientific theorizing. I am going to pursue this issue by way of two questions, one is an interpretative question and the other a textual one.

The interpretative question asks whether Wittgenstein's explicit metaphilosophy coheres with his actual method. I hope to make two points here. First, I would like to bring out the ways in which Wittgenstein's metaphilosophical remarks are more complicated than they are often presented as being. Second, I hope to show that some of the most emphasized passages do conflict with Wittgenstein's actual methods of argument. Wittgenstein's arguments are in the service of a theoretical diagnosis of philosophical error that relies on the careful structuring of Part I of the *Investigations.* This structure, I shall argue, provides an answer to the textual question: why do the metaphilosophical remarks occur where they do, that is, after a full examination of the relation of words to objects and before the rule-following considerations? Neither the constructivist nor the quietist has an answer to this question. The constructivist tends to ignore the structure of the *Investigations* in favor of finding whatever passages support the theory of language or mind that he finds in the text. The quietist, on the other hand, is more sensitive to the order of the passages and the importance of accommodating them all, but misses the content of Wittgenstein's theoretical diagnosis.

1 Wittgenstein's Explicit Metaphilosophy

Let me highlight the elements of a familiar sketch of Wittgenstein's metaphilosophy.[1] Wittgenstein's new philosophy is an activity, not a set of theses, doctrines, or theories. It is an activity that aims at showing that traditional philosophical theory building is 'a house of cards,' constructed with illusory materials in pursuit of ends which are themselves the result of confusion and mistaken understanding of the grammar of our language. The aim of the new philosophy is, thus, to bring philosophy to an end by disclosing these confusions and mistakes. The primary method for achieving this end is a method of description. Through the perspicuous display of timely reminders and descriptions of language in use, the theoretical and explanatory aspirations of traditional philosophy are revealed to overshoot the limits of language and so say nothing. The philosophical point of the descriptions and reminders is given by their relation to purported philosophical problems. Philosophy for the later Wittgenstein is thus wholly critical and reactive. It gets its point from the entanglement in confusion and illusion that results when individuals make reflective use of language to misguided ends. The metaphor of disease or mental illness is a crucial component of this sketch. Like neurosis, the only way to free oneself from the trap is by coming to realize that its problems and solutions are illusory, grounded in a willful and intellectualized imagination. There is nothing positive for philosophy to do, even though it will remain an ineradicable part of human life.[2] There is no question that the target of this dismal assessment is the scientistic optimism of the logical positivist movement in Anglo-American philosophy, the contemporary embodiment of perennial philosophical error.

Much in this thumbnail sketch is certainly to be found in Wittgenstein, both in his metaphilosophy and in his actual method. But if we have only this, the deep conflicts that the constructivist notes remain unassimilated. So, we need to look more closely at these passages. PI §§ 89–133 has three major stages. It opens with an initial contrast between the philosophical method of the Frege-Russell-*Tractatus* approach and his new method (PI §§89–91). The second stage characterizes the new method at work (PI §§92–120). Grammatical investigation aims at providing a theoretical diagnosis of how traditional philosophy goes wrong. These passages display a dialectical structure, in which an initial diagnosis of mistake (PI §§92–6) is followed by a closer examination of the phenomenology of philosophical puzzlement (§§97–103), which then permits a deeper diagnosis of philosophical error, one showing how the mistake and phenomenology of puzzlement connect (§§104–20). The target exemplifying this structure is the analytic method of Tractarian logical investigation. The final stage is a summing-up of the tools by which error can be removed (PI §§121–33).

Stage 1. Preliminaries: Logical Investigation vs. Grammatical Investigation (PI §§89–91)

The opening passages contrast the 'logical investigation' of contemporary philosophy that 'explores the nature of all things' (PI §89) with the new 'grammatical investigation' that 'sheds light on our problem by clearing misunderstanding away' (PI §90).[3] A logical investigation sees 'logic at the bottom of all the sciences' (PI §89); it is the science, as it were, of 'the "possibilities" of phenomena' (PI §90). As such, no empirical matters bear upon its success or failure. 'It is of the essence of our [logical] investigation,' Wittgenstein tells us, 'that we do not seek to learn anything *new* by it' (PI §89). A logical investigation seeks to give an account of the possibilities of phenomena in terms of the structure of thought. A grammatical investigation aims to show that this very goal is a mistake, one that can be cleared away by reminding ourselves 'of the kind of statement that we make about phenomena' (PI §90).

Demonstrating grammatical confusion is one of the motivations that inform logical investigation as well as Wittgenstein's new method. Central to the philosophical project of the *Tractatus* and logical positivism is the elimination of metaphysics as nonsense, by revealing misunderstandings concerning the use of words. 'Some misunderstandings', the later Wittgenstein agrees, 'can be removed by substituting one form of expression for another; this may be called an "analysis" of our forms of expression' (PI §90). Wittgenstein must distinguish his own new project, a grammatical investigation, from the analytic project of a logical investigation. Both purport to undermine certain forms of philosophical theorizing by showing how such theorizing is grounded in a misunderstanding of the grammar of our ordinary language. The difference, Wittgenstein tells us, is that in using analysis as the means for displaying misunderstandings, 'it may come to look as if there were something like a final analysis of our forms of language, and so a *single* completely resolved form of every expression' (PI §91). In this way, the method for revealing misunderstandings itself becomes the key component of the very philosophical project under attack. At the end of these passages, Wittgenstein will express the difference between logical and grammatical investigations by replacing the goal of 'complete exactness' (PI §91) with one 'aiming at ... *complete* clarity' (PI §133).

We see that in the opening passages Wittgenstein distinguishes his new grammatical investigation from the logical investigation exemplified in the *Tractatus* while acknowledging similarities that might obscure the key differences. For both methodologies, 'it is of the essence that we do not seek to learn anything new by it', that is, unlike science, no new empirical facts are to be uncovered (PI §89). Both seek to clear misunderstandings away by means of a philosophical investigation of language (PI §90). And both 'are trying to understand the essence of language' (PI §92). But, and here is the difference that makes all the difference, a grammatical investigation aims to show that the logical (or analytic) investigation itself is ensnared in the very misunderstandings that it purports to clarify. To show this requires considerable diagnostic work. Yet it is the key to Wittgenstein's critique of the *Tractatus*. Where the *Tractatus* sees confusion in metaphysical statements, the *Investigations* sees the same confusion at work in the method of analysis, and the explanatory work it assigns logic.[4] At the outset, then, we see that Wittgenstein's discussion of his own method and the way it differs from the analytic method of logical positivism is itself a diagnostic critique of the analytic method and the picture of language it holds in place.

Stage 2. Diagnosis: Misunderstandings and Bewitching Pictures (PI §§92–120)

The logical positivists and, arguably, the author of the *Tractatus* explain the mistakes of metaphysicians in terms of violating the logical form of language or failing to meet necessary conditions of meaningfulness. This positivist account of metaphysical error and the method of analysis that it relies upon are the object of critical scrutiny in these passages. Wittgenstein's commentary on his critique lays out the structure of his diagnostic argument, a structure to be found in his treatment of other philosophical theories. The diagnostic critique of the 'subliming of logic' is thus exemplary of his philosophical method. That diagnostic method is directed towards identifying the mistakes that engender and support a compelling picture of language and mind.

The commentary highlights three movements in the diagnostic process. Grammatical mistakes involving misunderstandings of 'the role of the ideal in our language' (PI §100) engender superstitious conviction (PI §110) that holds a picture in place for how things must be (PI §115), despite contradiction and anomaly, a picture that can be removed only by showing how it is self-defeating. In short, relatively shallow grammatical misunderstandings reveal deeper self-defeating bewitching pictures. These misunderstandings are grammatical, according to Wittgenstein, because they involve a misunderstanding of norms. The grammatical for Wittgenstein concerns the normative. Philosophical error originates in mistaking norms of linguistic practice for something else. Mistakes of this sort can look like discoveries from the perspective of a picture that has a non-rational grip on the philosopher. Critique of the picture requires more than the identification of misunderstandings. The picture itself must be shown to be completely inadequate for the explanatory task it is thought to achieve. This commentary is interwoven with an exemplifying critique of the method of analysis and the explanatory work assigned logic.

The new logic of Frege, Russell and the *Tractatus* is a powerful new notational system that provides an abstract representation of the functional structure of sentences and truth-functional relationships among sentences. The logical investigations of Russell and the *Tractatus* aim to reveal the misunderstandings of metaphysicians by using this new method of logical analysis. This method 'eliminates misunderstandings by making our expressions more exact' (PI §91). Wittgenstein has no objections to the limited and *ad hoc* application of the method of analysis. The mistake occurs when the philosopher thinks that the success of this method in some cases supports the idea that we can achieve 'a state of complete exactness' for all sentences. This eliding of the limited success of logical analysis with an ideal of complete exactness for any possible system of representation is the initiating mistake, according to Wittgenstein, one that becomes identified with a picture of language (or thought) as having a deep fully determinate structure. This ideal of complete exactness requires the uniqueness of thought, and this requirement itself, Wittgenstein argues, radically undermines the explanatory work that was to be done by thought. We find in PI §§92–9 a succinct critique that conforms to the three stages identified by the commentary. Grammatical misunderstanding is allied to commitment to a picture, which is radically self-defeating. Let me expand on this.

The ideal of complete exactness leads to seeing 'in the essence [of language], not something that already lies open to view and that becomes surveyable by a rearrangement, but something that lies *beneath* the surface' (PI §92). This misunderstanding of the *scope* of logical analysis, combined with the 'enormous importance attaching' to language, leads to a search for 'a pure intermediary between the propositional *signs* and the facts' (PI §94) – that intermediary is the thought or proposition. We make a further grammatical mistake on the way to transforming the new logical method into the a priori structure of language when 'we predicate of the thing what lies in the method of representing it' (PI §104). The *Tractatus* transforms a notational system by which to make certain features of language perspicuous into an idealizing notation that is identified with the deep structure of language itself, that is, with thought. We reach this conclusion because the ideal of complete exactness requires that language or thought has a unique logical structure. Propositions are introduced to be the bearers of deep logical form and to explain how language (a system of signs) says that things are thus and so in the world. This 'seduces us into thinking that something extraordinary, something unique, must be achieved by propositions' (PI §93). Propositions provide '... the unique correlate to reality' (PI §95). The impressiveness of the new logic now 'retreats', as Wittgenstein puts it, 'to these illusions' (PI §110). That thought is something unique 'proves to be a superstition (*not* a mistake!), itself produced by grammatical illusions' (PI §110). We now have the ingredients of a picture that holds us captive (PI §115). In understanding Wittgenstein's view, it is important to note that the superstition or picture that holds the philosopher captive is not itself a mistake. Expanding the scope of the method of logical analysis and attributing features of the notational system to what it represents are both mistakes. But the picture is fueled by something more potent than a scope fallacy or a confusion of metalanguage and object-language.

The picture to which we are now committed contains what Wittgenstein calls a paradox:

> 'Thought must be something unique.' When we say, and mean, that such-and-such is the case, we – and our meaning – do not stop anywhere short of the fact; but we mean: *this – is – so*. But this paradox (which has the form of a truism) can also be expressed in this way: Thought can be of what is not the case. (PI §95)

Just what is paradoxical about the claim that what we mean, that is, the thought expressed by the words we utter, is: this-is-so? Wittgenstein's suggestion is that thought as the unique correlate of reality cannot be of what is not actually the case. This is tantamount to holding that a thought or proposition cannot but be true. Yet this conflicts with an essential feature of meaningful propositions, namely, that they can be true or false without affecting their meaning. Propositions, as the unique intermediary between propositional signs and reality, explain the meaningfulness of the signs because they do not stop short of the facts. A proposition is what it is, in virtue of what is so, according to the proposition. This is what ensures the unique correlation of proposition to fact and how it is that a proposition has the very content it does. For example, the proposition that Baltimore has a harbor couldn't be the very proposition that it is without Baltimore's having a harbor. The meaningfulness or content of the proposition that Baltimore has a harbor is the fact that Baltimore has a harbor. With respect to this particular proposition, what-is-so is Baltimore's having a harbor. But this rules out the possibility of the proposition being the very proposition that it is and Baltimore's not having a harbor. The uniqueness of the proposition explains how language can say something about the world, but it does so only at the cost of not being able to say something that is contingently false. The bivalence of empirical sentences and their counterpart propositions is forfeited.

This unacceptable consequence can be avoided only if what makes a proposition that such-and-such the very proposition that it is, is not the fact that such-and-such. This requires that our meaning does stop short of the facts themselves. And now we need some intermediary between the proposition and reality, and an explanatory regress is initiated. Now, in effect, we need to identify the propositional content of the proposition. If what makes the proposition that Baltimore has a harbor the very proposition that it is, is not Baltimore's having a harbor, then it must be something else, say, the propositional content of the proposition. Thus, the explanatory work to be done by the proposition requires the proposition to be true. It looks like all thoughts must be true thoughts; otherwise they are meaningless. Yet this violates the bivalence condition that requires thought can be true or false.[5] Wittgenstein's claim is that the conception of thought as the unique correlate to reality, and so the necessary intermediary between mere linguistic signs and the facts, results in the impossibility of thinking (and so of saying meaningfully) what is not the case. Thus, a defining feature of language and thought, that it allows us to contemplate what is not actual, is rendered impossible by the ideal of complete logical analysis. I'll refer to this as the paradox of thought, and will return to it later.

To sum up this discussion, through misinterpreting our forms of language by identifying them with our idealizing representations, the problems that subsequently arise 'have the character of depth. They are deep disquietudes; their roots are as deep in us as the forms of our language and their significance is as great as the importance of our language' (PI §111). The importance and depth of philosophical problems are genuine. It is nothing short of the significance of language for our human form of life. To break free of the mistakes that produce the picture requires rejecting the picture, a picture that has become identified, by those bewitched, with our very form of life. In this way, the picture holds us captive (PI §115). The phenomenology of philosophical puzzlement is a crucial component of the philosophical project. The mark of being in the grip of a picture is a tolerance for contradiction and anomaly (PI §125). For, the disquietudes that the picture itself incurs are to be solved only with the resources of the picture. This is what explains the uncanny sense of depth that the emergence of contradiction can create.[6]

Stage 3. Summing Up: Philosophy as Therapy

The final set of passages (PI §§121–33) characterizes the method to be used in achieving a diagnosis that dispels picture driven theorizing. Here we have a description of philosophical perplexity from the outside as it were, as a kind of mental disturbance, characterized by disorientation with respect to the ordinary and familiar, for which therapy, not theory, is the only method that can free the philosopher from his perplexity.

> The philosopher's treatment of a question is like the treatment of an illness (PI §255).
>
> A philosopher is a man who has to cure many intellectual diseases in himself before he can arrive at the notions of common sense. (CV, p. 44, 1944)
>
> 'It is high time for us to compare these phenomena with something different' – one may say – I am thinking, e.g., of mental illness. (CV, p. 55, 1946)[7]

That therapy takes the form of a method of description, descriptions that provide 'a clear view of the use of our words...A perspicuous representation produces just that understanding which consists in "seeing connexions"' (PI §122). Clearly this passage alludes to the earlier diagnostic remarks concerning the subliming of logic and the mystifying of the proposition. We don't need to find a special object to mediate sign and use, but a way to array cases perspicuously such that we see the connections between words and their use. The point of describing such cases, whether found or invented, Wittgenstein emphasizes, derives from the philosophical problem being addressed. The method of description 'may in no way interfere with the actual use of language' (PI §124); it 'neither explains nor deduces anything' (PI §126). The aim is 'complete clarity' (PI §133). This is not the clarity of a totalizing theory, but a therapeutic clarity in which 'the philosophical problems should *completely* disappear' (PI §133).

These final summing up passages support the quietist reading. Wittgenstein's aim is to bring philosophical theorizing to an end by dissolving the problems they address through a description of our use of words (both found and invented). Success is achieved when we reach 'thoughts that are at peace. That's what someone who philosophizes yearns for' (CV, p. 43, 1944). The question for the next two sections is whether Wittgenstein's actual method of diagnosis and therapy conforms to these metaphilosophical remarks and achieve their happy end.

2 The Method of Description

Philosophy's proper task, then, is to dissolve philosophical problems through description, with an emphasis on the motley of language and the use of timely reminders and homely examples (cf. PI §§124, 127, 593). In doing so, it develops no theories or theses of its own (cf. PI §§124, 128, 599). These reminders and examples get their point in relation to the philosophical claim or picture they investigate. Yet Wittgenstein's actual reminders, timely examples, and descriptions of ordinary use seem inadequate to the task his metaphilosophy assigns them. Consider the following frequently cited cases.

Against the search for the essence of language, Wittgenstein asks 'But how many kinds of sentence are there? Say assertion, question, and command? – There are *countless* kinds.... And this multiplicity is not something fixed, given once for all...' (PI §23). Against denotational theories of meaning, Wittgenstein describes a simple grocers game (PI §1), though the actions of the buyer and the grocer hardly seem to describe what we do when buying apples. Against the ideal of determinate sense, Wittgenstein points out that the use of a word 'is not everywhere circumscribed by rules; but no more are there any rules for how high one throws the ball in tennis, or how hard; yet tennis is a game for all that and has rules too' (PI §68). And against the private ownership of experience, Wittgenstein reminds us of the following:

> Look at the blue of the sky and say to yourself 'How blue the sky is.' – When you do it spontaneously – without philosophical intentions – the idea never crosses your mind that this impression of colour belongs only to *you*. (PI §275)

By describing the grammar of various language games and the concepts implicit in them, we are to defuse the power that philosophical pictures exert. Yet these timely reminders and descriptions do very little to dissuade the philosopher from taking his theories seriously. Why should the problem of other minds dissolve at being instructed to look at a wriggling fly (PI §284) or reminded that one reacts immediately to the pain or fear of another (PI §303)? Why should the claim that sensations are private be repudiated because our everyday responses to colors take them to be public and shared? Why should the differences in the ways that the expressions 'five,' 'red,' and 'apples' are used convince us to give up the search for a theory of language? It is important to ask these questions since the method of description is supposed to free philosophers from pursuing these problems. Yet the method as characterized so far seems inadequate to its task. That task is to show that our ordinary linguistic practices are in order as they are and do not need philosophical buttressing or underpinning.

To make this point more sharply, let us compare Wittgenstein's descriptions of ordinary language with those provided by J. L. Austin who also uses a method of description to attack philosophical theories.[8]

In *Sense and Sensibilia*, for example, Austin gives nuanced descriptions of words typically used interchangeably by philosophers. His discussion (in Chapter 4) of the differences among 'looks,' 'seems,' and 'appears' quite effectively undermines A. J. Ayer's univocal use of these expressions. Ayer's theory of perception looks crude, his arguments fallacious, and his categorizations simplistic. This kind of careful description of English words seems to be just what Wittgenstein ought to be doing when he says that a grammatical investigation should describe our ordinary uses of words. Yet it is notable that this isn't what Wittgenstein does. He rarely gives such descriptions of ordinary usage. He does on occasion offer Austin-like descriptions, but they are not as convincingly done as Austin's. More to the point, they are not to be found in the *Investigations*, but in his manuscript musings about the differences between perception and imaging or the complexities of our concept of thinking. And, as we can see from the cases canvassed above, where he does offer descriptions, they are schematic, suggestive, concerned with what Stanley Cavell calls the 'generic object' (that is, objects that are stand-ins for any medium size physical object or any sensation or any perceptual state) rather than particular objects (such as glimpsing a finch rather than observing a warbler). His descriptions are closer to the examples and reminders of the sort used traditionally in philosophy to motivate a theory or a criticism. Indeed his engagement is at the same level of abstraction as the theories he targets for destruction. Wittgenstein's own particularism – that is, his emphasis on the plurality of cases that resists any form of essentialist categorization or explanation – looks to have a theoretical backing that does not reduce to descriptions of particular cases. One cannot but draw the conclusion that it is a mistake to try to assimilate Wittgenstein's method to that of ordinary language philosophy of the type exemplified by Austin. Yet the summing-up passages strongly suggest that this is just what Wittgenstein ought to be doing. The persuasive power of Wittgenstein's mode of argument does not lie with perspicuous nuanced descriptions of our ordinary expressions, but elsewhere.

If Wittgenstein's descriptions are not Austinian descriptions, how are they to be characterized? If we 'look and see,' we will find that they are of two kinds: the first are descriptions as reminders to the philosopher. They remind the philosopher that his theoretical claims conflict with what we all accept in the ordinary course of events. Such conflict is not an argument against the philosophical claim, but it is symptomatic of possible misunderstanding. Wittgenstein's distinctive mode of argument gets under way with the second kind of description he uses throughout Part I of the *Investigations*. These are his descriptions of language games. Indeed the language games he introduces exemplify what he has in mind by pursuing a method of description, and these language games are, for the most part, invented. They provide the medium in which the arguments of the *Investigations* are constructed.

There are six important language games in Part I of the *Investigations*. These are the builders (PI §2) and the extended builders game (PI §8); the simple color chart of PI §48; the standard meter stick in Paris (PI §50); the pupil learning the natural number sequence (PI §143); the private diary (PI §258); and the beetle in the box (PI §293).[9] Wittgenstein imposes two constraints on his construction of these language games. The first requirement is that the language game meets the requirements of the traditional philosophical theory under scrutiny, if any use of language does.[10] In other words, he introduces these cases as exemplary of the way language (or some aspect of language) is supposed to function according to some philosophical theory. The second constraint is that language games, though simplified and primitive, must involve recognizably ordinary uses of words. Implicitly (in his choice of games), he disallows the use of language games that can only be described using the technical vocabulary of the theory in question. Replacing philosophical ideals with ordinary cases (cf. PI §98) is his way of insisting that, if a philosophical theory of language is correct, we must be able to see how it explains ordinary uses of words. This concern is something that is strikingly missing in the *Tractatus*, where he cannot identify any elementary proposition or simple object except through the use of his proprietary vocabulary and the theory of which it is a part. So, he excludes philosophical language games from the arguments he constructs. This expresses his view that the burden of proof lies with the philosopher to show the relevance, for example, of 'being directly acquainted with sense data' to ostensive definition,[11] or 'asserting a proposition radical'[12] to meaning something, or 'applying the successor operator'[13] to learning or continuing the natural number sequence. Our interest now is in how these language games are used against traditional philosophical theories.

3 Wittgenstein's Distinctive Arguments: Mistake Leads to Paradox

Wittgenstein, like the great philosophers of the Western canon, introduces a new philosophical problem and a new argumentative strategy.[14] The problem is the problem of normative similarity, of how basic standards of sameness are set. The *Investigations* is structured around making this problem visible in four areas of intense philosophical interest: word-object relations and rule-following in the philosophy of language and the self-knowledge and ontology of sensations in the philosophy of mind. His new strategy uses two argument forms, each having origins as old as Plato, and each reworked in original and distinctive ways. These are the regress argument and the *reductio ad absurdum* argument. Philosophical problems originate in grammatical mistakes that occur when the philosopher misunderstands the role of the ideal in language (PI §100). In particular, the mistake is one of confusing the means of representation with what is represented (PI §104) We find this argument, which I shall call Wittgenstein's 'conflation argument', playing a central role in his discussion of the relation between words and objects in the first part of the *Investigations*. This replaces the positivist argument that the mistake of the metaphysician is to confuse the surface grammar of a sentence for its deep logical form.

The second Wittgensteinian argument occurs in four places in Part I of the *Investigations*. Three have had extensive discussion: the paradox of interpretation (PI §202), the private diary argument (PI §258), and the beetle in the box argument (PI §293). There is a fourth that occurs in the context of the metaphilosophical remarks themselves. This is the paradox of thought (PI §95) briefly discussed in Section 2 above. The paradox of thought reveals that the analytic method and the picture of language it supports is itself radically self-defeating. These arguments draw attention to what Wittgenstein calls paradoxes, and so I shall refer to them as Wittgenstein's 'paradox arguments'. The emergence of paradox marks philosophical bewitchment. This is the point at which the intolerable demands of the picture are made vivid and undeniable. Wittgenstein's argument is intended to show that the picture is radically self-defeating. This argument can be seen to replace the positivist account of metaphysical nonsense as that which violates verificationist strictures on meaning.

We need to understand these two argument forms, and importantly the relation they have to each other. Of the six methodologically central language games, three occur before the metaphilosophical remarks (the builders, the color chart, and the standard meter stick) and three occur after these remarks (the counting game, the private diary, and the beetle in the box). It's not the symmetry in number *per se* that I am emphasizing, but that the first three language games are all part of conflation arguments while the second three are implicated in paradox arguments. Conflation arguments reveal mistakes that initiate explanatory regress.

These mistakes involve, in some way, confusing the means of representation with what is being represented. Ideals are conflated with objects in ways that give rise to illusions of explanatory advance. The paradox arguments, on the other hand, reveal the special way in which theories guided by a bewitching picture are self-defeating: the theoretical explanations *eliminate* the very phenomena they are designed to explain. Both arguments are part of a theoretical diagnosis of how the philosophical theorist goes wrong and what motivates him. On Wittgenstein's diagnosis, grammatical mistake engenders the picture that takes on a life of its own. The picture promises an explanation of phenomena of the greatest importance, namely, of ourselves as thinking talking creatures. To break the grip of a bad picture requires showing that the promise of explanation is illusory. To align these two arguments with the diagnostic sketch of the metaphilosophy, we need to see that the conflation arguments are aimed at identifying grammatical mistakes and the paradox arguments are directed against bewitching pictures.

First, the conflation arguments. What is it to confuse the means of representation for what is represented? This conflation, Wittgenstein thinks, is the initiating mistake of representationalist theories of language. It involves treating an exemplary object (an ideal), which can be used to explain or justify the application of a term, as a metaphysically special object to which the term applies in some especially strong way. Regress arguments purport to show that such explanations fail because the explanans reproduces, as part of the explanation, the very feature(s) it was to explain. Aristotle's Third Man Argument against Plato is the classic example. Aristotle points out that the problematic relation that unites all particular men in a common class is the same relation that Plato holds obtains between all particular men and the universal form Man, and is equally problematic. Wittgenstein's conflation arguments are related to explanatory regress arguments, but focus on the mistake that makes such an apparently explanatory move tempting. Where Aristotle aims to show the inadequacy of the explanation, Wittgenstein would look to understand why Plato thought that the appeal to the universal form Man was explanatory in the first place.

The first three language games each exemplify a philosophical thesis that gains its plausibility from an unrecognized conflation of normative and non-normative features – a misunderstanding of the ideal in language. The builders game models language in accord with a representationalist picture, according to which a sentence represents a state of affairs in virtue of the naming relation that obtains between the individual words and objects. Though commentators typically construe the builders game as directed against denotational theories of meaning, it in fact plays a role in two conflation arguments, the target being Frege's two semantic properties (sense and reference). This is, in effect, Wittgenstein's critique of the primacy of assertoric form thesis. The sense of a sentence – that things are thus-and-so – and the reference of its non-logical constituents in virtue of which it has the sense it has, require the primacy of assertoric form in all meaningful uses of language. And this requires drawing a distinction between the sense of a sentence and the use to which it is put, a distinction that is essential to all language use and thought. This primacy thesis is a mistake, according to Wittgenstein, one initiated by conflating a notational device (Frege's assertion sign) with linguistic action (entertaining and asserting a propositional content) (PI §22).

The second conflation argument involving the builders is the more familiar argument, one directed against the idea that ostensive definition can fix reference *ab initio*. The denotational theory of meaning results from conflating ostensive definition (using an object as an exemplar in the explanation of the meaning of a word) and ostensive teaching (causally effecting an association between a word and a kind of object through training into a practice) (PI §6). The simple color chart of PI §48 is used to show that the philosophical ideal of absolute simplicity involves conflating our normal reactions to the elements of the chart with properties intrinsic to those elements. The distinction between simple and complex is relative to our interests and our normal reactions, where both are normatively structured. But the distinction is mistakenly taken to reflect categorical properties of the objects themselves. The standard meter stick (PI §50) is used to examine the thesis that there must be some ultimate simple substance of the world that necessarily exists as the standard of sameness. This involves taking the role of the meter stick as a standard for categorical properties of the metal bar itself. All of these are cases in which the means of representation is conflated with what it represents.[15]

Conflation arguments identify the misunderstandings that open the way to misplaced and misconceived theorizing about the normativity of language. For each of the conflation arguments Wittgenstein develops against the representationalist theory of language, he constructs a complementary paradox argument. The argument against the primacy of assertion is completed with the paradox of thought; the argument against ostensive definition as fixing meaning is tied to the paradox of interpretation; the argument against absolutely simple objects can be paired with the private diary argument (the paradox of self-knowledge);

and the argument against simple objects as the substance of the world is paired with the beetle in the box argument (the paradox of sensation). Wittgenstein takes the paradox arguments and the philosophical theories they target to strike directly at the picture that allows conflation to look like discovery. And for each pair we see a thesis about what is essential to language finding its completion in a thesis about the nature of mind.

Wittgenstein's paradox arguments reveal the contradictory character of philosophical theories. In another context, Wittgenstein suggests that a contradiction in 'a calculation made for a practical purpose... might be conceived as a hint from the gods that I am to act and *not* consider' (RFM IV.56, p. 254). Wittgenstein takes this hint in his own treatment of the paradoxes that emerge within philosophical theories of language and mind. They are *reductio ad absurdum* arguments of a special sort. Wittgenstein gives these arguments a special twist that makes them his own. The contradictions that mark these as paradoxes emerge in the same way in each of the arguments. First, just as language games are constructed that best fit the philosophical theory under scrutiny, so the philosophical theories themselves seem, *prima facie*, the best able to explain the phenomena under inquiry. Sentences say something about the world because they express thoughts that provide the unique correlate to reality. Rules determine correct applications because they guide and justify continuations within a game. Self-knowledge is incorrigible because of the subject's privileged relation to his own mental events. The interiority of the mind is best explained by the existence of private mental objects. Yet, and this is the second feature, these theories that seem best able to explain the phenomena under inquiry result in the elimination of the very phenomena they are to explain. The mechanics of the explanation itself eliminate its target. This is what makes these theories peculiarly self-defeating, and worthy of the name 'paradox'. It is not just a matter of finding inconsistency within the theory, but that the explanatory features of the theory itself eliminate the phenomena the theory is tailored to explain.

The paradox of thought discussed above has just this structure. Thought (or propositions) seem best suited to explain how language says something substantive about the world (that is, something that can be true or false) because thought has the constituent structure to connect it to languages (the logical form of propositions mirrors the logico-syntactic structure of sentences) and the content to connect it directly to reality (that things are thus-and-so). But paradoxically how the appeal to thought or propositions explains the meaningful use of substantive empirical sentences, eliminates *empirical* sentences. If empirical sentences are meaningful sentences that can be either true or false, then there are no empirical sentences. The contrast between being true or false and being necessarily true collapses. This is the paradox of thought.[16]

There is a similar structure to the other paradox arguments. Interpretation seems best suited to explain how we use a rule or formula to go on in the same way, for it selects one method of projection among the many possible ones. But paradoxically it is interpretation that obliterates the very distinction between correct and incorrect, and so the intelligibility of going on in the same way. The contrast between correct and incorrect collapses. This is the paradox of interpretation. Merely having a sensation should set the standard for recognition (or self-knowledge), but the hypothesis that the sensation itself sets the standard renders us mute: 'one gets to the point where one would like just to emit an inarticulate sound' (PI §261). What is 'easiest to know' in Descartes' phrase (namely, the contents of one's own mind) is rendered unknowable. Cartesian transparent subjectivity eliminates self-knowledge. The contrast between is and seems collapses. This is the paradox of self-knowledge. And the appeal to sensations as private objects eliminates sensations altogether. The contrast between something and nothing collapses. This is the paradox of consciousness. Wittgenstein's point, in all four cases, is that these philosophical theories of meaning, of rule-following, of self-knowledge, and of sensation, all eliminate just what they were introduced to explain, namely, the semantic use of truth (and falsity), the distinction between correct and incorrect continuations, the special character of subjective knowledge, and the ontology of sensation (or consciousness more generally) in our lives. These are the paradoxes created by theories of language and mind.

It can be objected that the distinction I am drawing between conflation arguments and paradox arguments is overstated.[17] There is, for example, a way of looking at the critique of ostensive definition that sees this argument as also eliminating the very phenomena it was to explain, namely, meaning. Wittgenstein objects that ostensive definition cannot fix the meanings of terms *ab initio* because it presupposes the meaningful use of language. The converse of this is that if ostensive definition were the only way to fix meaning, meaning would be eliminated. Thus, the allegedly special character of the paradox arguments can be found in the conflation arguments occurring before the metaphilosophical remarks. It is true that at a sufficiently abstract level, one can describe both argument forms as showing that philosophical theories of meaning and the ontological commitments they support eliminate meaning. But this does not show that the two arguments are the same, nor that we have reached the same point in the theoretical diagnosis. Conflation arguments identify grammatical misunderstandings, that is, various ways in which normative features of language are treated as categorical features of some thing. These arguments invite the response that we have been looking in the wrong place for understanding norms, not that the project of finding an explanation in terms of some special objects is misguided *tout court*. The positivists show how agreement with the conclusions, as it were, of conflation arguments, namely, that grammatical misunderstandings have occurred, doesn't free one from philosophical illusion. More sophisticated strategies identify logic and mind as explanatorily fundamental. Paradox arguments, on the other hand, are all directed towards showing that fundamental normative distinctions collapse given the philosophical theories that articulate the picture that seems so compelling. This difference, I shall argue, fits Wittgenstein's diagnosis of philosophical error in terms of grammatical mistakes that engender and support pictures that compel assent.

4 Conflict and Agreement between Metaphilosophy and Method

Now let's return to the opening questions of this chapter. Why do the metaphilosophical passages occur where they do? Are these passages located here for some methodologically sound reason, or could they just as well have occurred elsewhere, say, at the beginning of the *Investigations* or the end or interpolated in some other location? Why not, for example, at the end of the rule-following discussion and before the private language argument, where Wittgenstein shifts his attention from language to mind? And how far does Wittgenstein's actual method agree with his explicit metaphilosophy? The quietist reading of the therapeutic interpretation privileges the metaphilosophical remarks, especially those disavowing any form of theorizing and affirming commitment to a method of description only. The constructivist reading sees a deep conflict between the metaphilosophy and the alternative theory of language and mind purportedly to be found in the *Investigations*. This is a false conflict. Both sides give a disproportionate importance to the final summing-up passages of the metaphilosophical remarks.

The summing-up passages support the quietist interpretation. They state that the proper aim of philosophizing is to bring philosophical theory-building to an end by using a method of description by which we achieve a complete clarity with respect to the grammar of language. But as I have shown, this sketch of Wittgenstein's method is incorrect or at least highly misleading. His descriptions are not those that are part of an Austinian or ordinary language philosophical project. They are part of a more complex argumentative methodology that employs conflation and paradox arguments as stages in an overall strategy for undermining the source of philosophical theories of language and mind.

The quietist must develop a more sophisticated defense of her strongly therapeutic interpretation. She can urge, correctly, that Wittgenstein, in showing the paradoxical character of philosophical theories, does not endorse any philosophical conclusions, for example, that meaning is indeterminate or interpretation is inherently unstable. Nor does Wittgenstein endorse any claim to the effect there is nothing special about avowals or that there are no sensations. These negative conclusions are the consequences of the theories Wittgenstein rejects. Thus, it could be argued, it would seem, that if these two argumentative strategies – the conflation argument and the paradox argument – were the whole story of Wittgenstein's actual method, this would support the quietist reading of the later philosophy. Problems internal to philosophical theories result in their implosion. Nothing positive from the outside is involved in their destruction. In self-destructing, they leave the terrain of our ordinary language untouched and in no need of philosophical comment. But the sophisticated quietist must admit that there is conflict between the metaphilosophical remarks and the method of argument. As we have seen, Wittgenstein does not use a method of description only. (Or it requires a very elastic notion of what counts as description, so elastic as to be of little use in contrasting Wittgenstein's method from other methods). But the conflict between method and metaphilosophy runs deeper than this.

First and most importantly, there is need for further diagnosis of why and how philosophical theorizing goes wrong. Why is the reification of logic a mistake rather than a discovery? What further diagnosis does Wittgenstein offer that allows the philosopher to make the intellectual shift from 'discovery' of simple objects or logical form to mistake? As we have seen, living with contradiction is a mark of being in the grip of a philosophical picture.[18] The full diagnosis of the power of this picture comes with recognizing the depth and pervasiveness of the problem of normative similarity, and the significance of Wittgenstein's own treatment of our judgments of normative similarity. Wittgenstein's theoretical diagnosis is rooted in an alternative picture of language and mind, one centering on how to understand our bedrock judgments of normative similarity. It is a picture that can only be seen through the elimination of the philosophical pictures that have obscured our actual form of life.

Second, Wittgenstein's longing for complete clarity achieved through a perspicuous surview of language is itself an illusion. Indeed it is the expression of the very conception of philosophy Wittgenstein seeks to overcome. His metaphilosophy replaces the ideal of a complete and final theory of language with the ideal of complete clarity. But his actual argumentative and diagnostic arguments make complete clarity as unobtainable as a complete theory. The final conflict between metaphilosophy and method comes with the very success that Wittgenstein has in undermining the pictures that inform philosophical theorizing. This success opens the door to other issues and questions, centering on normativity, necessity, and certainty, rather than representation, reality, and knowledge. The philosophical problematic is not eliminated, it is changed.

Yet against constructivist interpretations, Wittgenstein does not offer a replacement theory of language or mind to rival that of Russell or Carnap. Certainly, relative to the totalizing theories of language and mind the logical positivists present, Wittgenstein does not offer a theory of either. Language is not a totality, having the recursive formal structure of a logical language. Analysis, as a method for revealing the implicit structure of propositions, is at best a tool that is effective in an *ad hoc* and piecemeal fashion. And there is no terminus of ultimate analysis for sentences although our bedrock judgments of sameness, for example, 'This is red' or '1, 2, 3…' have been mistaken for such a terminus. In his rejection of a scientistic conception of theory building, there is complete harmony between his metaphilosophy and his actual method. Others have made this point, and so I won't pursue it any further.

Rather the point I want to emphasize is that there is agreement between his actual method and the first two stages of his metaphilosophical comments. Relative to the theorizing of a logical investigation, Wittgenstein's grammatical investigation is descriptive. But Wittgensteinian reminders of ordinary practice and descriptions of language games are part of his more general argumentative and diagnostic strategy. They are most effective when used at the completion of a diagnostic argument. When the space is cleared, we can recognize the import of these reminders. Further, Wittgenstein's actual method fits the diagnostic sketch he offers in the second stage of his metaphilosophical remarks. Wittgenstein's very understanding of philosophical perplexity and theorizing involves the philosopher's move from grammatical misunderstanding to being in the grip of a picture. The misunderstandings are identified by conflation arguments and the theories sustained by commitment to a picture are revealed to be paradoxically self-defeating.

This takes us to our textual question. Why do the metaphilosophical remarks occur where they do? Wittgenstein's account of philosophical entrapment provides the reason. They occur at the point at which the *Investigations* shifts from the conflation arguments directed against grammatical mistakes to the paradox arguments directed against theories. The remarks prepare the reader for this deeper diagnosis of philosophical illusion.

Notes

* Earlier versions of this chapter have been given at the Wittgenstein Workshop at The University of Chicago, the annual meeting of Mind & Society at the University of Manchester, and the Wittgenstein and Twentieth Century Philosophy Conference at Beijing University. I want to thank the participants at these conferences for their helpful remarks and criticisms.

1. I shall be drawing primarily on two sources for my discussion of Wittgenstein's metaphilosophy. These are *Investigations* §§89–133 and various passages from *Culture and Value*. There are, however, a number of other important metaphilosophical remarks scattered throughout Part I: §§254, 255, 309, 371, 373, 392, 464, 593, 599. We find interpreters as different as P.M.S. Hacker and Marie McGinn endorsing the basic elements of this sketch. See Hacker (1972, Chapter V) and McGinn (1997, Chapter 1). Also see my (1988).
2. In *Culture and Value*, Wittgenstein says '...we are still occupied with the same philosophical problems as were the Greeks...it has to be so. It is because our language has remained the same and keeps seducing us into asking the same questions' (Wittgenstein, 1980, p. 15).
3. Though Wittgenstein uses 'logical' often interchangeably with 'grammatical' or 'conceptual,' here it is clear that he is contrasting a logical investigation with a grammatical one.
4. This is a matter of considerable importance in coming to understand the *Tractatus* correctly. It comports with much contemporary interpretation of that early work that emphasizes the new logic as what drives the philosophical work done. This in contrast to the idea that the pivot on which the argument of the *Tractatus* turns is the search for a theory of representation. See Ricketts (1996).
5. Charles Travis has developed a similar interpretation of these passages on the subliming of logic, but he approaches the issue from a different direction. He puts the conflict in terms of two different ways of counting propositions, one formal and the other semantic. These two ways of counting propositions can

come apart in ways that vitiate the explanatory work to be done by propositions. See his Chapter 3, 'Hardness,' (unpublished manuscript).

6. In these metaphilosophical passages of the *Investigations*, Wittgenstein is describing his own bewitchment in the *Tractatus*. The picture that held him captive there is the picture of language as the unique correlate to reality. The very fact that this picture leads to paradox is exemplary of the power that philosophical pictures exert over those who become enthralled. The austere reader takes the paradox as evidence that Wittgenstein could not have endorsed the picture that leads to the paradox. I think, on the contrary, for the early Wittgenstein paradox was indicative of the profundity of the picture. This is what Wittgenstein is describing in the metaphilosophical remarks we are now discussing.
7. It is important to note that the two passages from *Culture and Value* were written in 1944 and 1946. Though the imagery of illness is strong in Wittgenstein, he doesn't introduce this imagery until the 1940s. A clear change from the early to the later Wittgenstein is the association of philosophizing with mental illness. The *Tractatus* metaphilosophy is directed to an intellectualized or task oriented methodology, and with that an intellectual explanation of philosophical error. The error is the mistaken belief that philosophy is a science, and so the arena of its own problems, phenomena, and theories. The limits of the sayable put philosophy outside the cognitive arena. The later philosophy retains the view that traditional philosophy mistakenly aims at being a science, at constructing theories, following a 'scientific' methodology, and the like. But the diagnosis of the later philosophy identifies the motivation not with the intellect but with the volition: 'What has to be overcome is a difficulty having to do with the will, rather than the intellect' (CV, p. 17). It is an intellectualizing willing that things be a certain way. When such a willing becomes disconnected from 'any true or false beliefs about natural processes' (CV, p. 44), it becomes pathological. It is arguable that in the very late writings (1949–51), there is a further shift in Wittgenstein's orientation, one that involves a different assignment for mental illness. Now Wittgenstein tends to apply it to those incapable of being trained into our (human) bedrock practices. We treat such people as suffering from a mental disturbance or mental illness. And Wittgenstein distinguishes the philosopher from this condition. Again, just as philosophy is not science (though philosophers will it to be such), so philosophical puzzlement is not mental illness (though philosophers sound like the mentally disturbed).
8. Stanley Cavell in his (1979, Chapter III) brought to our attention important differences between Wittgenstein and Austin. Of particular interest is his discussion of the contrast between the generic object of traditional philosophical interest and the particular object of Austin's descriptions.
9. There are other examples of language games as well but they do not play the important methodological role that these six cases do. These include the grocers game (PI §1) and the reading game (PI §156) as well as several others. There are language games listed as evidence of the multiplicity of language games (PI §23) as well as passages concerned with the language games of using 'true' and 'false' (PI §136), lying (PI §249), expressing sensations (PI §§288 and 290), and so on. None of these play the sustained argumentative role that the methodological ones do.

10. David Stern makes this point as well. See his (2004, Chapter I).
11. (Russell, 1912, Chapter V); and (Russell, 1956).
12. (Frege, 1980).
13. (Frege, 1964); (Wittgenstein, 1922, 4.1252).
14. The great philosophers are notable for introducing new questions and constructing new ways of arguing. The new modes of argument themselves create a recognition of the problem as well as the means for solving it. The Socratic method both reveals the inadequacies of the commonsense reliance on particular examples and shows what is required to address this defect, namely, to identify the essences that unite the particulars. Descartes' method of doubt shows the singular importance of answering the opening question of the *Meditations* – what are the firm and constant foundations of knowledge? – as well as providing the key for answering it. Kant's critical philosophy looks for the conditions necessary for the possibility of experience, and constructs the transcendental deduction as a solution. Each new question is tied to a potent new form of argumentation. Wittgenstein has also achieved this.
15. For a fuller discussion of the last three conflation arguments, see Williams (1999).
16. I have only given a sketch of what I am calling the paradox of thought. This argument has not been developed and widely discussed as is true of the other three paradox arguments. It deserves treatment in a separate paper in which the conflation argument that culminates in PI §22 and the paradox argument of PI §95 are fully explored and related.
17. I want to thank David Stern and Paul Boghossian for raising this objection.
18. Jose Saramago (2002) captures this most aptly in his philosophical novel. He writes of the relation between contradiction and coherence as follows: 'A person or character contradicts himself within the bounds of his own inner coherence, whereas incoherence, which, far more than contradiction, is a constant behavioral characteristic, resists contradiction, eliminates it, cannot stand to live with it. From this point of view, and at the risk of falling into the paralyzing webs of paradox, we should not exclude the hypothesis that contradiction is, in fact, one of the most coherent contraries of incoherence' (Saramago, 2002, p. 186). The paradox of the *Tractatus* is a contradiction within the bounds of its own inner coherence though the austere reader of the *Tractatus* would hold that it is full incoherence only. For a full treatment of the austere reading of the *Tractatus*, see my (2004).

Part IV
Concluding Remarks

12
Farewell Thoughts on Wittgenstein

Anthony Kenny

In this concluding chapter, I intend to develop four themes of Wittgenstein's philosophy in the light of the discussions of the other contributors – sometimes in agreement, sometimes in disagreement. The four topics I have chosen are: verification and metaphysics; mind and behaviour; rules and conventions; method and metaphilosophy.

1 Verification and metaphysics

Like the logical positivists Wittgenstein saw the removal, the dissolution, of metaphysics as one of the principal tasks of the philosopher. But unlike the positivists, he did not think this could be achieved by the use of a single, universal, philosophical sledgehammer such as the verification principle. Bill Child has well brought out the complicated nature of the mature Wittgenstein's attitude to verificationism. Already in 1958 Peter Geach used the example of Queen Victoria's deathbed in order to distance Wittgenstein from positivism.

> Wittgenstein did not want to deny the obvious truth that people have a 'private' mental life, in the sense that they have for example thoughts they do not utter and pains they do not show; nor did he try to analyse away this truth in a neo-behaviouristic fashion. In one of his lectures he mentioned Lytton Strachey's imaginative description of Queen Victoria's dying thoughts. He expressly repudiated the view that such a description is meaningless because 'unverifiable'; it has meaning, he said, but only through its connexion with a wider, public, 'language-game' of describing people's thoughts. (Geach, 1958, p. 3)

Child is right to stress that Wittgenstein continued to believe that there was an important relation between meaning and verification: but as is said in PI §353, which Child quotes, asking whether and how a proposition can be verified is only one particular way of asking 'how d'you mean?' Verification is only a single element in the language games which confer meaning. Language contains commands as well as statements, and expresses wishes as well as beliefs. In describing the language game with the word 'pain' we must take account not only of third-party statements about pain, but also of first-person expressions of pain. Moreover, as this example shows, it is a mistake to take whole propositions, rather than individual expressions, as the basic units of meaning. We cannot ignore the composite nature of sentences, which enables us to present new thoughts with old words. To be sure, we cannot combine just any old expressions to make sensible sentences: we can give no meaning to 'it is five o'clock on the sun.' But our very knowledge that this sentence is senseless derives from our awareness of the use of its component parts.

Verification itself, for the later Wittgenstein, is not a simple matter. As is well known, he made a distinction between two kinds of evidence that we may have for the obtaining of states of affairs, namely *symptoms* and *criteria*. Where the connection between a certain kind of evidence and the conclusion drawn from it is a matter of empirical discovery, through theory and induction, the evidence may be called a *symptom* of the state of affairs; where the relation between evidence and conclusion is not something discovered by empirical investigation, but is something that must be grasped by anyone who possesses the concept of the state of affairs in question, then the evidence is not a mere symptom, but a *criterion* of the event in question. Such is the relation between scratching and itching, between crying and pain.

Exploiting the notion of *criterion* enabled Wittgenstein to avoid Cartesian dualism without falling into behaviourism. He could agree with dualists that particular mental events could occur without accompanying bodily behaviour; on the other hand he agreed with behaviourists that the possibility of ascribing mental acts to people depends on such acts having, in general, a behavioural expression.

Cartesianism is a metaphysical theory in the sense of isolating statements about mental life from any possibility of verification or falsification in the public world. But Cartesianism was far from being the only form of metaphysics that Wittgenstein criticised. The kind that he most systematically attacks is the metaphysics that consists of grammar masquerading as science. One source of metaphysics is the philosopher's temptation to mimic the claims and methods of science. Metaphysics, in this sense, is a quasi physics, an imaginary physics elevated into something sublime and mysterious. In the paper he presented at the conference from which this volume originates, Fred Stoutland emphasised Wittgenstein's hostility to scientistic metaphysics and went on to say, about the tendency of philosophers to answer questions in the way science does, 'The most vulgar examples of these tendencies we find, it seems to me, in contemporary philosophy of mind, be it in the form of the physicalist's identification of so called mental states with brain processes or the eliminative materialists rejection of our common sense psychological concepts – what they call folk-psychology.'[1]

When we consider human thought the metaphysical impulse may lead us to postulate spiritual substances, or spiritual processes. When grammar makes us expect a physical substance, but there is not one, we invent a metaphysical substance; where it makes us expect an empirical process, but we cannot find one, we postulate an incorporeal process. As examples of metaphysical substances we might think of the Cartesian ego, or the homunculus that often seems to creep into popular scientific explanations of the operations of the brain. Wittgenstein's treatment can expose the absurdity of the notion of such metaphysical substances but he devotes more energy and patience into the task of disabusing us of metaphysical processes.

A paradigm of the antimetaphysical therapeutic process is the section on reading and reading machines which is so meticulously analysed in Joachim Schulte's paper. In this section, Wittgenstein tells us, he is 'not counting the understanding of what is read as part of "reading" for purposes of this investigation: reading is here the activity of rendering out loud what is written or printed; and also of writing from dictation, writing out something printed, playing from a score and so on'. (PI §156)[2] Wittgenstein chose this rather arcane example for discussion, I believe, because it is the simplest possible case in which the temptation arises to postulate metaphysical mental processes. His treatment of the issue consists of a series of dialogues with three metaphysical interlocutors, whom we might call Empiricist, Mentalist, and Materialist respectively. Empiricist thinks that reading must involve a special conscious experience; Mentalist thinks it must be an unconscious process of a hidden mental machinery, Materialist thinks that it must be the setting up of a special connection set up in the brain. Wittgenstein brings out the incoherence of each of these metaphysical positions in the case of simple reading, and then goes on to generalise his critique in the much more complicated case of understanding, with all the semantic elements that that involves. He patiently dismantles the three possible versions of the idea that understanding is a mental process, in parallel to his dissolution of the metaphysical accounts of reading.

In fact, understanding cannot be thought of as a process at all. Understanding is a kind of ability, and is therefore a state rather than a process – but it is not a state of some metaphysical ego or mind, but of the human being whose behaviour provides the evidence for the possession of the relevant abilities. Wittgenstein disposes of metaphysics in the philosophy of mind not by the blunt instrument of some positivistic verification principle, but by the careful drawing of distinctions that enable him to disentangle the mixture of truism and nonsense in the metaphysician's muddle.

2 Mind and behaviour

Important though it is in his philosophy, hostility to metaphysics is only one element of Wittgenstein's thought. The kind of metaphysics that is exemplified by empiricists, mentalists, and materialists is certainly, and rightly, rejected by Wittgenstein. But if by metaphysics we mean the kind of thing that goes on in Aristotle's work of that name, then Wittgenstein was no mean practitioner of the discipline and nowhere more so than in his philosophy of mind. The distinction between actuality and potentiality, and the classification of different kinds of potentiality, is universally recognised (by both friend and foe) as one of Aristotle's most characteristic contributions to philosophy and in particular to philosophy of mind.

Among the powers that are peculiar to human beings, according to Aristotle, there are some that are innate, such as the sense-faculties, and other, like the ability to play the flute, are acquired by practice. The liberal arts, and in general the skills that are the fruits of education, are powers of a particular kind, namely abilities. Abilities have exercises which are the relevant scientific, artistic, and craft activities; but abilities are themselves actualisations of the human capacity to learn that is presupposed by education. Aristotelians called them first actualities, and their exercises second actualities.

Aristotle's distinctions, and the scholastic developments of them, were mocked as gobbledygook by philosophers in the early modern period. But Wittgenstein devoted the greater part of two books (*The Blue Book* and *The Brown Book*) to the discussion of different kinds of actuality and potentiality, and he employs the distinctions he draws to great effect in discussing human thought and behaviour. His philosophy of mind is thus, to a great extent, Aristotelian.

'Only of a living human being' Wittgenstein wrote 'and what resembles (behaves like) a living human being can one say: it has sensations; it sees; is blind; hears; is deaf; is conscious or unconscious' (PI §281). This does not mean that Wittgenstein is a behaviourist: he is not identifying experience with behaviour, or even with dispositions to behave. The point is that what experiences one can have depends on how one *can* behave. Only someone who can play chess can feel the desire to castle; only a being that can discriminate between light and darkness can have visual experiences.

The relation between experiences of certain kinds, and the ability to behave in certain ways, is not a merely contingent connection. Mental and emotional states have primitive, natural expressions, as well as more complicated learned expressions (such as their expression in language). These expressions are criteria for the states they express. It was this notion of criterion, as mentioned above, that enabled Wittgenstein to steer between the Scylla of dualism and the Charybdis of behaviourism.

If it is wrong to identify the mind with behaviour, it is even more mistaken, according to Wittgenstein, to identify the mind with the brain. Such materialism is in fact a grosser philosophical error than behaviourism because the connection between mind and behaviour if a more intimate one than that between mind and brain. The relation between mind and behaviour is a criterial one, something prior to experience; the connection between mind and brain is a contingent one, discoverable by empirical science. Any discovery of links between mind and brain must take as its starting point the everyday concepts we use in describing the mind, concepts which are grafted on to behavioural criteria.

Oddly enough, developments in the philosophy of mind since Wittgenstein have shown that it is possible to combine the errors of materialism with those of dualism. One of the standard dualist misunderstandings of the nature of the mind is the picture of mind's relation to body as that between a little person or homunculus on the one hand, and a tool or instrument or machine on the other. This misunderstanding is compounded if we assign the role of homunculus to the brain, identifying it, as materialists do, with the mind so conceived.

What is wrong with the homunculus fallacy? In itself there is nothing misguided in speaking of images in the brain, if one means patterns in the brain, observable to a neurophysiologist, that can be mapped onto features of the visible environment. What is misleading is to take these mappings as representations, to regard them as visible to the mind, and to say that seeing consists in the mind's perception of these images. The misleading aspect is that such an account pretends to explain seeing, but the explanation reproduces exactly the puzzling features it was supposed to explain.

Peter Hacker and Max Bennett prefer to call the fallacy 'the mereological fallacy': they point out that it is a fallacy that can be committed in connection with a mechanism like a clock no less than in connection with an animal such as a human being.

> We called the mistake 'mereological' because it involves ascribing to parts attributes that can intelligibly be ascribed only to the wholes of which they are parts. A form of this error was pointed out around 350 BC by Aristotle, who remarked that 'to say that the soul psuchè is angry is as if one were to say that the soul weaves or builds. For it is surely better not to say that the soul pities, learns or thinks, but that a man does these with its soul.' (De An. 408b12–15)...Our primary concern was with the neuroscientific cousin of this, namely the error of ascribing to the brain – a part of an animal – attributes that can be ascribed literally only to the animal as a whole... In Aristotelian spirit we now observe that to say that the brain is angry is as if one were to say that the brain weaves or builds. For it is surely better to say not that the brain pities, learns or thinks, but that a man does these. Accordingly, we deny that it makes sense to say that the brain is conscious, feels sensations, perceives, thinks, knows or wants anything – for these are attributes of animals, not of their brains. (Bennett and Hacker, 2007, p. 131–2)

The identification of mind and brain is a category mistake, because the brain is a material object and the mind is a capacity. The mind is the capacity to acquire intellectual abilities. The possessor of human mental capacities is neither the mind nor the brain but the human being.

Wittgenstein, like Aristotle, worked hard to bring home to his readers the crucial importance of the distinction between potentiality and actuality, both in general and in the particular case of human thought and knowledge. In order to illustrate the stratification of different levels of potentiality, Aristotle's favourite example is the very same one as Wittgenstein used in the passage about reading that is illuminatingly discussed in Joachim Schulte's paper. Aristotle points out that we have first an ability to learn Greek, which is a power or *dynamis* (one of the powers of the soul), and that this potentiality is actualised when we learn the language. But this actualisation is only a first actualisation, which is itself a potentiality. The potentiality is exercised, Aristotle says, when the student identifies a particular letter as an alpha. This shows that Aristotle's *grammatike* is the ability to read in the very basic sense in which Wittgenstein is treating of it. Just as Wittgenstein discusses such things as whether there is a first moment at which someone is reading, so Aristotle, in treating of the *hexeis* of the virtues in the *Nicomachean Ethics* discusses whether it is possible to do just acts without yet being just.

Reflection on the nature of abilities enables us to answer the question, raised by a number of contributors, whether thought has priority over language or language has priority over thought. Some of those who argue for the priority of thought insist that there can be thoughts without language, as in the case of animals; and Alberto Voltolini points out that Wittgenstein countenanced this possibility when he spoke of a dog being afraid that his master would beat him (PI §650). The key to this issue is that only those can have thoughts who can express thoughts, and that a thought is identified as the thought it is by what counts as its expression.

A dog, when he sees his master take the lead off the hook on the door, thinks that he is going to be taken for a walk and expresses the thought in a very clear manner by leaping and bounding about and pawing at the door. But only very simple thoughts have a similar, non-linguistic, expression. We cannot imagine such an expression for the thought that truth is beauty, or that the stars are many light years away. Most thoughts that are worth thinking are thoughts that only a language-user can have, because only in language could they be expressed. And language has priority over thought in such cases, because an ability is defined in terms of its exercise, not vice versa.

The distinction between actuality and potentiality is neglected equally by Cartesian dualists and by scientistic materialists. But to avoid confusion in this area it is necessary to introduce a further concept, not developed either in Aristotle or Wittgenstein. This is the notion of the *vehicle* of a capacity: that is to say, the actuality that experience shows must belong to the capacity's possessor if it is to exercise its capacity. Uncontentious examples are: the shape of a key (the vehicle of its capacity to open the door), the alcohol in whisky (the vehicle of its capacity to intoxicate). The notions of possessor, capacity, exercise, and vehicle all need to be applied in order to achieve clarity in thinking about the human mind: for the relation of the brain to the mind is precisely that of vehicle to capacity.

The only thing that is wrong with Wittgenstein's philosophy of mind, in my view, is that he is insufficiently explicit about the relationship between capacities and their vehicles. But his silence here may not be inadvertence; for in a notorious passage in *Zettel* he shows himself willing to countenance the idea that capacities (e.g., the capacity to recognise someone after many years of absence) may exist without having any material vehicle (§610). He is undoubtedly right that there is nothing conceptually incoherent in the supposition; but I share the queasiness about this passage that is widespread even among W's admirers. Much of history suggests that scientific progress is made by inquiring into the vehicles of capacities, rather than by treating capacities such as dormitive powers as the basic building blocks of science. Wittgenstein's suggestion seems to amount, in Aristotelian terms, to suggesting that there may be formal and final causes in the absence of efficient and material ones. Perhaps there can: but one cannot help worrying whether Wittgenstein may here be adopting towards contemporary neuroscientists the attitude that renaissance Aristotelians adopted to Galileo.

3 Rules and conventions

Games differ from each other by having different rules: rugby differs from soccer because the rules of rugby differ from those of soccer and chess differs from draughts because the rules of the two games are not the same. The same goes for language games. A language game is a set of rules for the use of words: in that sense language is normative. As Han-jo Glock says, in using a word I commit myself to a rule of use; just as I do if I move a piece in chess or make a serve in tennis. But what kind of rules are we talking about, and what is it to follow such a rule?

The essential element in the normativity of language is that there should be a difference between a right and a wrong use of a word, and the minimum requirement for one to be following a rule is that one should be aware of this difference. But the commitment to a rule may be, and commonly is, implicit, and the rule-follower may well be quite unable to articulate the rule being followed.

Horwich offers an illuminating discussion of implicit rule following. For S to be following a rule R implicitly, he says, S's activity must be governed by the ideal law R. He cites Wittgenstein's remark in PI §54 that an observer can read the rules of a game from the way it is played in practice of the game, like a natural law governing the play. There will commonly, Horwich says, be some tendency for S to correct instances of nonconformity. But self-correction, he goes on to argue, plays no part in the explanation of the word's overall use; hence we should call into question the whole idea that meaning is rule-following. The operation of the 'natural law' does not have to be a case of rule-following in order to engender meaning.

This account seems to me an inadequate account of the normativity of language. As said above, there has to be a difference between right and wrong use, and the language user has to be aware of this difference. Horwich denies that we know by introspection the rules that we are following. In this he is right, because two different kinds of knowledge are in question. There is the practical knowledge that we possess and exhibit in our actual observance of the rule and our correction of error: this is not any object of introspection, but is something that comes out in behaviour. On the other hand there is the theoretical knowledge that permits an explicit articulation of the rule in question: this we normally do not possess, unless we are professional linguists. (The 'rules' that are formulated by philosophers, such as 'Call x "a dog" iff x is a dog' are of course useless for any practical purpose.)

There is, however, a major defect in Horwich's 'covering law' analysis in that it ignores the huge difference between obeying a natural law and following a rule. The rule-following behaviour of language users and game players, unlike the behaviour of physical objects conforming to gravity and inertia, is a series of voluntary intentional actions. If specifying the meaning of a word was simply giving the law that describes the actual behaviour of language users, then the law would have to be written in such a way as to include the mistakes and failures of use no less than correct and successful use. Being rule-governed is being different from being governed by causal laws. Rules differ from such laws in several ways. Rules may be broken; not so causal laws, short of a miracle. To be governed by a rule you must be conscious of it at least to the extent that you are aware of any breach of it. But there is no need for you to be aware of being governed by a causal law: the planets know nothing of Kepler's laws.

If the use of language consists of a series of actions aiming to conform to norms, what kinds of norms are in question? Clearly, they are not Kantian categorical imperatives. On the other hand I do not think that they are merely permissive norms (as Glock suggests) nor desire-generated imperatives (as some followers of Davidson claim). Rather, they are imperatives derived from the structure of the activity one is participating in. Here one may call to mind Wittgenstein's *Lecture on Ethics*.

> Supposing that I could play tennis and one of you saw me playing and said 'Well, you play pretty badly' and suppose I answered 'I know I'm playing badly but I don't want to play any better' all the other man could say would be 'Ah then, that's all right.' But suppose I had told one of you a preposterous lie and he came up to me and said 'You're behaving like a beast' and then I were to say 'I know I behave badly, but then I don't want to behave any better', could he then say 'Ah, then that's all right'? Certainly not. (LE, pp. 38–9)

Whether or not I want to be any good at tennis, if I serve into the net twice, that is a double fault; the rules of tennis are not merely permissive. On the other hand, as Wittgenstein points out there is no rule about how high you throw the ball when serving, and that does not prevent tennis from being a game with rules. So too with language: there are rules laid down in the language games, but not everything is determined by rules, so that it is a mistake to regard language as a deterministic calculus. (cf. PI §68)

Glock maintains that the norms of language are not only the rules but conventions: that is to say, they are arbitrary rules that are shared with others. In support of this one might quote PI §202 'it is not possible to obey a rule "privately".' But against Glock Horwich claims that when Wittgenstein says it is not possible to obey a rule privately 'the alternative to "*private* rule following" that Wittgenstein is insisting upon is "rule following *as a practice*" – and that can perfectly well be constituted in the regular activity of a single isolated person.'[3] I think Horwich is in the right here. Certainly the private language argument does not rule out the possibility of an isolated individual devising a language. The private language that Wittgenstein demolished is one whose words refer to 'what can only be known to the person speaking: to his immediate private sensations' (PI §243) – or to similarly private entities such as Cartesian thoughts, Lockean ideas, Humean impressions, Russellian sense-data. The private language that is Wittgenstein's target is not concerned with the items in the public environment of isolated individuals. We can imagine Robinson Crusoe keeping a record of the weather for himself, by marking on a notched stick symbols that he has invented for sun and cloud and rain. Two things distinguish such a language from the mythical private language: first Crusoe can correct himself (waking up one morning and finding massive puddles about, he may delete the 'sun' sign he absent-mindedly carved the day before); second, other humans, if they land on the island, can quickly be brought to understand the system of signs. If Glock is to be saying that the norms of language must be conventions, he must in his definition of convention as arbitrary rules replace 'shared with others' with 'shareable with others'.

But though a nonshared language-game is possible, few such games would have much point and it is clear as a matter of fact that the overwhelming majority of language games do involve communication and therefore shared conventions. Hence arises a puzzle which has concerned a number of our contributors, and which is well formulated by Glock. If the meaning of an expression *e* in a language L is determined or constituted by the use speakers make of *e*, the meaning of *e* cannot at the same time prescribe a certain use to the speakers of L. Glock's answer to this problem seems to me adequate and decisive: *communal* use *constitutes* meaning, while *individual* use is *answerable* to it. Let me illustrate this point by recounting a relevant experience of my own.

I once received an invitation from Oxford University Press to attend a party to celebrate the publication of one of their authoritative dictionaries of the English language. I was puzzled by the invitation, as I had had no part in the preparation of the dictionary. When I arrived at the party, I discovered that all the other invitees were similarly puzzled. The reason for our invitation became clear only when each of us was given a parting gift: the list of the quotations from our individual works that had been cited in the dictionary as instances of current English usage. It was in our role as quoted authorities that we had received our invitations.

My first reaction was of immense pride: what an honour, I thought, to be treated by such an august institution as a paradigm user of the English language. My second reaction was rather different. If the rules of English usage, it occurred to me, are determined only by people like me and my fellow guests, why should I pay any attention to those rules? I felt rather like Groucho Marx when he said that he did not want to belong to any club that was willing to have the likes of him as a member. Later, reflecting on Kant's notion that we moral agents are all equal legislators in the kingdom of ends, I realised that the OUP party was an illustration of the truth that all users of language are equal legislators in the republic of words.

If we accept that language is essentially normative, is the same true of thought? Voltolini claims that though language is normative, thought is not, because it is only complex thoughts that require language for articulation. I take it that he would agree that the thoughts that are expressible only in language must be answerable to norms no less than language itself is. But even the simple thoughts of animals, if they are to be thoughts at all, must be capable of being mistaken. The happy dog, leaping about in expectation of his exercise, may be disappointed when he finds his master has taken the lead off the hook only in order to repair the strap. But where there is right and wrong – in this case true and false – there is normativity.

Some philosophers believe that language ability is innate, and therefore language cannot be a system of conventions. Thus Horwich writes 'Insofar as we allow that ... mental terms may turn out to be innate and universal (i.e., symbols of Mentalese) then the idea that meaning is rule following becomes highly conjectural'.[4] In this context it is important to distinguish between the origin of a skill and its operation. Let us suppose that the ability to use language does have an innate element: none the less, what the ability consists in is the ability to operate a set of conventions. Wittgenstein, in the course of the private language argument, draws precisely this distinction between the origin and operation of a skill. If human beings showed no signs of pain, he observes, a child could not be taught the meanings of words. 'Well' he goes on 'let's assume the child is a genius and itself invents a name for the sensation!' (PI §257) The private language argument is not about the acquisition but about the exercise of meaning. What the exercise of meaning consists in is a matter of philosophy; what elements are necessary for the ability to be acquired is a matter for empirical investigation.

Some philosophers wonder how the philosophy of language can be, as Wittgenstein says it is, a descriptive activity if language is itself normative. I would myself endorse the response of Glock, who says '"In English, it is correct to apply 'drake' to an object x iff x is a male duck" states a fact about the behaviour of Anglophones. But it does so only because the Anglophone speech community accepts a certain norm, a standard of correctness.'[5] Though languages are systems of norms, they do not escape being part of our natural history. This is repeatedly emphasised by Wittgenstein. 'Commanding, questioning, recounting, chatting, are as much a part of our natural history as walking, eating, drinking, playing.' (PI §25) 'What we are supplying are really remarks on the natural history of human beings; we are not contributing curiosities, however, but observations which no one had doubted, but which have escaped remark only because they are always before our eyes.' (PI §415)

4 Method and metaphilosophy

Once upon a time it was easy to classify propositions. There were analytic propositions, true in virtue of their meaning, and there were synthetic propositions that contained information about the world. All analytic propositions were known a priori; empirical propositions, known a posteriori, were all synthetic. Out of respect for Kant philosophers would inquire whether any propositions were synthetic a priori but commonly the question was raised only to be answered in the negative.

In the latter part of the twentieth century the tidy dichotomy – analytic a priori vs. synthetic a posteriori – came under attack from two opposite directions.

In 1951 Quine's 'Two Dogmas of Empiricism' sought to undermine the analytic/synthetic distinction, replacing it with a continuum of propositions of varying degrees of entrenchment in the web of our beliefs. Quine denied that analyticity could be defined in terms of synonymy or necessity, and he rejected the idea that a sentence was synthetic if it could be verified or falsified by experience. It is not single sentences, he argued, but whole systems – which include mathematics and logic as well as geography and history – that are verified or falsified. 'Our statements about the external world face the tribunal of sense experience not individually but only as a corporate body…A conflict with experience at the periphery occasions readjustments in the interior of the field. Truth values have to be redistributed over some of our statements.' (Quine, 1953, pp. 41–2) We cannot single out a class of analytic statements which remain true whatever happens; no statement – not even a law of logic – can be totally immune to revision. This position of Quine was, in essence, accepted also by Donald Davidson.

In extreme contrast to this position Saul Kripke, as Diego Marconi reminds us, contended that there was a class of propositions that were both a posteriori and necessary. Whereas for Quine no proposition was necessary, not even laws of logic, for Kripke there were some necessary facts that were discovered by empirical inquiry. These two contrasting philosophies, however, had one thing in common: they called in question the distinction between science and philosophy. This placed them both, in their opposite corners, at odd with Wittgenstein. For Wittgenstein science and philosophy are two totally different activities, science concerned with information and explanation, philosophy with description and understanding. For Quine, Davidson and Kripke, on the other hand, there is a continuum between philosophy and science, between physics and metaphysics, between empirical psychology and philosophy of mind.

It must be admitted that it is difficult to reconcile Wittgenstein's metaphilosophical statements about the nature of philosophy with his actual practice. On the one hand he tells us that philosophy is purely descriptive: it neither explains nor proves anything, and it only states what everyone admits. Any genuine proposition must be capable of being true or being false, and hence there are no necessary propositions. There are no arguments in philosophy, and no conclusions: the philosopher's task is simply to assemble reminders of the obvious. On the other hand, the *Philosophical Investigations* appears to contain passages of tough argumentation, and to draw from these passages conclusions such as that there is no such thing as a private language. It also is full of statements such as 'The meaning of a word is its use in the language' and 'essence is expressed by grammar' that appear to be philosophical theses that are by no means obvious and that not everyone will admit.

In fact, Wittgenstein's metaphilosophical remarks give an accurate, but incomplete, account of Wittgenstein's general philosophical practice at the time when he was working on the *Investigations*. This can be illustrated from the discussion of thought and thinking that begins at section 318. There, Wittgenstein explores and explodes a number of philosophical errors about the nature of thinking. One error attacked is the idea that thought is a mental process accompanying spoken sentences. 'Thinking', Wittgenstein says, 'is not an incorporeal process which lends life and sense to speaking' (PI §339). Is this not a philosophical thesis, the conclusion of an argument? Wittgenstein at once rejects this interpretation of his remark. 'Am I acquainted with incorporeal processes, then, only thinking is not one of them? No; I called the expression "an incorporeal process" to my aid in my embarrassment when I was trying to explain the meaning of the word "thinking" in a primitive way.' (*ibid.*)

'Thinking is an incorporeal process' is not a false statement, an error to be denied. The sentence can even be used to express, in a clumsy and misleading way, recognition of the difference between the grammar of 'eat' and 'think'. What is rejected is a pseudoproposition, the expression of a misleading imagination. When, in another passage, Wittgenstein rejects talk of an inner process of remembering, he is not denying anything, only setting his face against a futile philosophical picture (cf. PI §306).

If Wittgenstein is successful he persuades the reader to join him in rejecting the idea that thinking and remembering are inner, incorporeal, processes. It is not by offering reasons that he persuades, though it is rational, not unreasonable, for the reader to accept his persuasion. Wittgenstein proceeds not by presenting arguments for a negative conclusion, but by assembling reminders of the obvious. These reminders include the philosophical statements that, once uttered, are assured of universal assent. But they are also of several other types: many of them are not statements at all, but questions, or commands, or jokes.

It is, indeed, remarkable how little of Wittgenstein's text consists of statements of any kind. If we take, as a sample chosen more or less at random, sections 501–30 of the PI, we find that they contain 105 sentences. Less than half of these (43) are in the indicative mood at all: 35 sentences are questions, 17 are quotations (sentences for discussion) and 10 are commands (usually to carry out a thought-experiment). Of the indicative sentences many simply set the stage for an example, or expand upon targeted quotations.

What, then, are the theses in philosophy that Wittgenstein has in mind, the statements that everyone admits? I suggest the following as examples.

> Cheese does not grow or shrink without a cause. (cf. PI §142)
> Dogs do not talk to themselves. (cf. PI §357)
> If I say falsely that something is red, then it isn't red. (cf. PI §429)
> Nothing could induce me to put my hand into a flame.
> (cf. PI §472)
> A smiling mouth smiles only in a human face. (cf. PI §583)
> When I raise my arm, my arm goes up. (cf. PI §621)

These are all truisms that are, on the face of it, devoid of philosophical content. What makes them philosophical is that they are presented in the course of the treatment of a philosophical problem. That they are being used as philosophical may often be brought out by an unusual emphasis. Thus, the sentence in §429 is best translated 'If I say falsely that something isn't red, then all the same it is *red* that it isn't.'

In spite of Wittgenstein's claim that philosophy only states what everyone admits, such truisms of everyday are not the most important method that Wittgenstein uses in the clearing up of philosophical confusion. A technique much more often employed is the drawing of distinctions. If someone says that sensations are private, the first step in enlightening him is to propose two different paraphrases of 'private', in PI §246 and §253. To clarify the concept of thinking, we distinguish between instantaneous understanding (PI §§319–26), thoughtful utterance (PI §§327–39), talking in one's head (PI §§344–8), and so on. Someone who wants to say something metaphysical about thought may well be cured of his wish by being reminded of the many different ways in which 'think' is used. Again, if we want to get clear about an emotion, like fear, we need to make a distinction between its object and its cause (PI §476). And again, we can understand a sentence in two very different senses of 'understand' (PI §531).

Another favoured method of achieving philosophical clarity is to draw a comparison between different human activities that provide the context of language games, so that the learner is assisted in seeing the similarities and the differences between them. Thus, 'Understanding a sentence is much more akin to understanding a theme in music than one may think' (PI §527). For this purpose, of course, Wittgenstein often invents imaginary language-games. These are not approximations to an explanation of language, but '*objects of comparison* which are meant to throw light on the facts of our language by way not only of similarities, but also of dissimilarities' (PI §130).

Philosophical confusion, we might say, happens when the imagination takes over the role of the intellect. We have a picture of how a word is used, and the picture conflicts with our understanding of the word, which is expressed in our actual use of it. Wittgenstein's method of dealing with this is to give the imagination enough rope to hang itself. The vague and confused and ultimately nonsensical picture we have is to be painted in concrete detail, which will make its nonsense patent. The anti-private-language 'argument' is a paradigm of this kind of treatment.

In what sense is it true that there is only description in philosophy and that there are no philosophical theories? Meredith Williams points out that there are two different forms of description offered by Wittgenstein. In one form he reminds a philosopher that his claims conflict with what we normally do. But in other instances the language games he describes are invented.

In discussing dreaming, Wittgenstein describes an actual language game (cf. RPP, I §368; PI, II, vii, p. 184). People narrate something to us after waking up; we thereupon teach them the expression 'I dreamt' followed by the narrative. I then sometimes ask them: 'Did you dream anything last night?' and sometimes get an affirmative, sometimes a negative answer, sometimes none. That (Wittgenstein says) is the language game. More interesting and more common is the case where Wittgenstein describes a would-be or putative language game as part of a *reductio ad absurdum*. This is a very different kind of description.

Wittgenstein's procedure in such cases has very well described by Williams. A philosophical theory is introduced that seems the best able to explain a set of phenomena under discussion. A language game is devised to illustrate the way language is supposed to function according to the theory under scrutiny. It is constructed without the use of the proprietary vocabulary of that theory. The reader is then led to realise that the theory itself results in the elimination of the very phenomena it was meant to explain. Thus Wittgenstein shows that Cartesian transparent subjectivity eliminates self-knowledge and the appeal to sensations as private objects eliminates sensations altogether.

As I have argued elsewhere,[6] I do not believe that it is in the end possible to reconcile Wittgenstein's account of philosophy with the entirety of his philosophical activity in the *Investigations*. Others will perhaps be more successful in reconciling his theory with his practice. Those who find it impossible to make the reconciliation will be forced in the end to make a choice between accepting his theory and following his practice. If one gives priority to the theory, then perhaps the simplest method of effecting some kind of reconciliation is to look on the substantive theses of the *Investigations* as momentary yielding to a form of philosophical expression which he had long rejected in theory, and to a great extent grown out of in practice. The sentences in question are, of course, among the most quoted ones of the *Investigations*. None the less, it might be claimed, nothing would be lost if they were all excised from its text, and there would be a substantial gain in consistency.

On the other hand, one might give priority to the practice, and write off the theory. The metaphilosophy, it might be claimed, is an inadequate account even of Wittgenstein's own philosophy: it is more obviously inadequate as an account of the best practice of the philosophers of the past. It is no surprise if Wittgenstein's account of the nature of philosophy should turn out to be one of the weakest parts of his philosophising: the same is true of the greatest philosophers from Plato and Aristotle onwards. If we discard Wittgenstein's account, we can treat philosophical statements, in accordance with tradition, as being bearers of truth values: as being either necessary true or necessarily false. We can then treat philosophical arguments, including Wittgenstein's own, as perfectly genuine arguments proceeding in accordance with normal logical rules.

For my part, I find it hard to decide between these alternatives; sometimes I am tempted by the one, and sometimes by the other. I wish to conclude not by arguing in either direction, but by drawing attention to the fact that Wittgenstein's own thought on the nature of philosophy was fluid, and did not remain fixed in the pattern of the remarks in the *Big Typescript* and in the *Investigations*. Shortly before his death, when writing the notebooks that became *On Certainty*, he became more and more interested in propositions of the kind that I listed earlier as reminders of the obvious, and less concerned to make the grammatical generalisations that are difficult to reconcile with his second-order remarks. But the tension remained to the end.

Such statements as 'humans have brains in their skulls' and 'the earth existed long before my birth' and 'if someone's arm is cut off it will not grow again' began to exercise him as occupying a middle ground between grammatical propositions and empirical propositions. In OC he asks what the relation is between these certain propositions, propositions that 'stand fast' for us, and the language-games that give meaning to our utterances. He replies:

> Propositions of the form of empirical propositions, and not only propositions of logic, form the foundation of all operating with thoughts (with language). (OC §401)

He goes on immediately:

> In this remark the expression 'propositions of the form of empirical propositions' is itself thoroughly bad. (OC §402)

He could not hide from himself that there were more things in his philosophy than could be confined within his metaphilosophy.

Notes

1. Stoutland, this volume, p. 46.
2. Cf. Schulte, this volume, p. 33.
3. Horwich, this volume, p. 133.
4. Horwich, this volume, p. 126.
5. Glock, this volume, p. 97.
6. Cf. Kenny (2008, pp. 147–8).

References

Wittgenstein's texts

Wittgenstein's Lectures, Cambridge 1932–1935 (AWL), A. Ambrose (ed.) (Oxford: Blackwell) 1979.

The Blue and Brown Books: Preliminary Studies for the 'Philosophical Investigations' (BB) R. Rhees (ed.) (Oxford: Blackwell) 1958.

The Big Typescript (BT) (Wien: Springer-Verlag) (*Wiener Ausgabe*, vol. XI), 2000; ed. and tr. by G.G. Luckhardt and M.A.E. Aue (Oxford: Blackwell) 2005.

Culture and Value (CV) G.H. von Wright and H. Nyman (eds.) (Oxford: Blackwell) 1980.

Eine philosophische Betrachtung (EPB), *Werkausgabe* Band 5 (Suhrkamp: Frankfurt am Main) 1984, 117–282.

'A Lecture on Ethics' (LE), in J.C. Klagge and A. Nordmann (eds.) *Philosophical Occasions 1912–1951* (Indianapolis, IN: Hackett) 1993, 37–44.

Wittgenstein's Lectures on Philosophical Psychology 1946–47 (LPP), P. Geach (ed.) (London: Harvester) 1988.

Last Writings on the Philosophy of Psychology vol. II (LW II), G.H. von Wright and H. Nyman (eds.) (Oxford: Blackwell) 1992.

Wittgenstein's Lectures, Cambridge 1930–1932 (LWL), D. Lee (ed.) (Oxford: Blackwell) 1980.

Notebooks 1914–1916 (NB) (London: Routledge and Kegan Paul) 1961.

On Certainty (OC), G.E.M. Anscombe and G.H. von Wright (eds.) (Oxford: Blackwell) 1969.

Philosophical Grammar (PG) R. Rhees (ed.) (Oxford: Blackwell) 1969.

Philosophical Investigations (PI), G.E.M. Anscombe and R. Rhees (eds.) (Oxford: Blackwell), second edition, 1958.

Philosophische Untersuchungen: Kritisch-genetische Edition (PU), J. Schulte (ed.) in collaboration with H. Nyman, E. von Savigny and G.H. von Wright (Frankfurt am Main: Suhrkamp) 2001.

Remarks on Colour (RC), G.E.M. Anscombe, L. McAlister, and M. Schättle (eds.) (Oxford: Blackwell) 1977.

Remarks on the Foundations of Mathematics (RFM) G.H. Von Wright, R. Rhees, G.E.M. Anscombe (eds.) (Oxford: Blackwell), third edition, 1978.

Remarks on the Philosophy of Psychology vol. I (RPP I), G.E.M. Anscombe and G.H. von Wright (eds.) (Oxford: Blackwell) 1980.

Remarks on the Philosophy of Psychology vol. II (RPP II), G.H. von Wright and H. Nyman (eds.) (Oxford: Blackwell) 1980.

Tractatus logico-philosophicus (TLP) (London: Kegan Paul, Trench, Trubner and Co.) 1922; D. Pears and B.F. McGuinness (tr.) (London and New York: Routledge) 1961 rev. edn. 1974.

Zettel (Z), G.E.M. Anscombe and G.H. von Wright (eds.) (Oxford: Blackwell) 1981.

Other texts

G.E.M. Anscombe (1957) *Intention* (Oxford: Blackwell).
—— (1981) *Metaphysics and Philosophy of Mind* (Minneapolis, MN: University of Minnesota Press).
J.L. Austin (1962) *How to Do Things with Words* (Oxford: Clarendon Press).
K. Bach (1988) 'Burge's New Thought Experiment: Back to the Drawing Room', *Journal of Philosophy*, vol. 85, 88–97.
J. Bacon (2002) 'Tropes', in E.N. Zalta (ed.), *Stanford Encyclopedia of Philosophy*, http://plato.stanford.edu/archives/fall2001/entries/tropes/
G.P. Baker and P.M.S. Hacker (1980) *Wittgenstein. Understanding and Meaning, An Analytical Commentary to the Philosophical Investigations*, vol. 1 (Oxford: Blackwell).
—— (1984) *Language, Sense and Nonsense* (Oxford: Blackwell).
—— (1985) *Wittgenstein. Rules, Grammar and Necessity, An Analytical Commentary on the Philosophical Investigations*, vol. 2 (Oxford: Blackwell).
M.R. Bennett and P.M.S. Hacker (2007), 'The Conceptual Presuppositions of Cognitive Neuroscience: A Reply to Critics' in M.Bennett, D. Dennett, P. Hacker, J. Searle (eds), *Neuroscience and Philosophy* (New York: Columbia University Press), 127–62.
G. Bergmann (1964) *Logic and Reality* (Madison, WI: University of Wisconsin Press).
A. Bilgrami (1993) 'Norms and Meaning', in R. Stoecker (ed.), *Reflecting Davidson* (Berlin: de Gruyter), 121–44.
P. Bloom (2002) *How Children Learn the Meanings of Words* (Cambridge, MA: MIT Press).
P.A. Boghossian (2002a) 'How are Objective Epistemic Reasons Possible?', in J. Bermudez and A. Miller (eds.), *Reason and Nature* (Oxford: Oxford University Press), 1–47.
—— (2002b) 'The Rule-Following Considerations', in A. Miller and C. Wright (eds.), *Rule-Following and Meaning* (Chesham: Acumen), 141–81.
—— (2003a) 'Blind Reasoning', *The Aristotelian Society*, suppl. vol. 77, 225–48.
—— (2003b) 'The Normativity of Content', *Philosophical Issues*, vol. 13, 33–45.
O.K. Bouwsma (1986) *Wittgenstein's Conversations 1949–1951* (Indianapolis, IN: Hackett).
J. Brandl (1996) 'Intrinsische und abgeleitete Intentionalität', in A. Schramm (hrsg.), *Philosophie in Österreich 1996* (Wien: Hölder-Pichler-Tempsky), 187–99.
R. Brandom (1994) *Making It Explicit* (Cambridge, MA: Harvard University Press).
—— (2000) *Articulating Reasons* (Cambridge Mass: Harvard University Press).
T. Burge (1979) 'Individualism and the Mental', *Midwest Studies in Philosophy*, vol. 4, 73–121.
—— (1986) 'Intellectual Norms and Foundations of Mind', *Journal of Philosophy*, vol. 83, 697–720.
C. Calabi and A. Voltolini (2005) 'Should Pride of Place be Given to the Norms?', *Facta Philosophica*, vol. 7, 85–98.
H. Cappelen and E. Lepore (2005) *Insensitive Semantics* (Oxford: Blackwell).
R. Carnap (1947) *Meaning and Necessity* (Chicago, IL: University of Chicago Press).

S. Cavell (1962) 'The Availability of Wittgenstein's Later Philosophy', *Philosophical Review*, vol. 71, 67–93.
—— (1979) *The Claim of Reason* (Oxford: Oxford University Press).
D. Chalmers (1996) *The Conscious Mind* (Oxford: Oxford University Press).
D. Chalmers (2002a) 'On Sense and Intension', in J. Tomberlin (ed.), *Philosophical Perspectives 16: Language and Mind* (Oxford: Blackwell), 135–82.
—— (2002b) 'Does Conceivability Entail Possibility?', in T. Szabò Gendler and J. Hawthorne (eds.) (Oxford: Clarendon Press), 145–200.
N. Chomsky (1988) *Language and Problems of Knowledge* (Cambridge MA: MIT Press).
A. Coffer (1991) *The Semantic Tradition from Kant to Carnap: To the Vienna Station* (Cambridge: Cambridge University Press).
A. Coliva (ed.) (2009) *Essays in Honour of Crispin Wright* (Oxford: Oxford University Press).
D. Davidson (1980) *Essays on Actions and Events* (Oxford: Oxford University Press).
—— (1984) *Inquiries into Truth and Interpretation* (Oxford: Oxford University Press).
—— (1985) 'Replies to Essays', in B. Verrazano and M. Hintikka (eds.), *Essays on Davidson: Actions and Events* (Oxford: Oxford University Press), 195–252.
—— (1986) 'A Nice Derangement of Epitaphs', in E. Lepore (ed.), *Truth and Interpretation* (Oxford: Blackwell), 433–46.
—— (1990) 'The Structure and Content of Truth', *Journal of Philosophy*, vol. 87, 279–328.
—— (1994a) 'Replies', in R. Stoecker (ed.), *Reflecting Davidson* (Berlin and New York: de Gruyter).
—— (1994b) 'The Social Aspect of Language', in B. McGuinness and G. Oliveri (eds.), *The Philosophy of Michael Dummett* (Dordrecht: Kluwer), 109–26.
—— (1995) 'Locating Literary Language', in R.W. Dasenbrook (ed.), *Literary Theory after Davidson* (University Park, IL: Pennsylvania State University Press), 295–307.
—— (1997) 'Seeing Through Language', in J. Preston (ed.), *Thought and Language* (Cambridge: Cambridge University Press), 15–27.
—— (1999a) 'The Centrality of Truth', in J. Peregrin (ed.), *Truth and Its Nature (if any)* (Dordrecht: Kluwer), 105–15.
—— (1999b) 'Replies', in L.E. Hahn (ed.), *The Philosophy of Donald Davidson* (La Salle: Open Court).
—— (2001) *Subjective, Intersubjective, Objective* (Oxford: Oxford University Press).
—— (2004) *Problems of Rationality* (Oxford: Oxford University Press).
—— (2005) *Truth, Language, and History* (Oxford: Oxford University Press).
D.C. Dennett (1987) 'Evolution, Error and Intentionality', in *The Intentional Stance* (Cambridge, MA: The MIT Press), 287–321.
—— (1990) 'The Myth of Original Intentionality', in K.A. Mohyeldin Said *et al.* (eds.), *Modelling the Mind* (Oxford: Clarendon Press), 43–62.
M. De Caro, F. Ferretti, and M. Marraffa (eds.) (2007) *Cartographies of the Mind* (Dordrecht: Kluwer).
F. Dretske (1988) *Explaining Behavior* (Cambridge, MA: The MIT Press).
—— (1990) 'Misrepresentation', in W.G. Lycan (ed.), *Mind and Cognition* (Oxford: Blackwell), 129–43.
—— (1995) *Naturalizing the Mind* (Cambridge, MA: The MIT Press).

M. Dummett (1973) *Frege. Philosophy of Language* (London: Duckworth).
—— (1981) *The Interpretation of Frege's Philosophy* (London: Duckworth).
M. Dummett (1986) 'A Nice Derangement of Epitaphs: Some Comments on Davidson and Hacking', in E. LePore (ed.), *Truth and Interpretation* (Oxford: Blackwell), 459–76.
—— (1993) *The Seas of Language* (Oxford: Oxford University Press).
—— (1994) 'Reply to Davidson', in B. McGuinness and G. Oliveri (eds.), *The Philosophy of Michael Dummett* (Dordrecht: Kluwer), 257–62.
M. Eklund (2002) 'Inconsistent Languages', *Philosophy and Phenomenological Research*, vol. 64, 251–75.
R. Elugardo (1993) 'Burge on Content', *Philosophy and Phenomenological Research*, vol. 53, 367–84.
G. Evans (1982) *The Varieties of Reference* (Oxford: Clarendon Press).
H. Field (1992) 'Physicalism', in J. Earman (ed.), *Inference, Explanation, and Other Frustrations* (Berkeley, CA: University of California Press), 271–91.
M.O. Fiocco (2007) 'Conceivability, Imagination and Modal Knowledge', *Philosophy and Phenomenological Research*, vol. 76, 364–80.
J.A. Fodor (1987) *Psychosemantics* (Cambridge, MA: The MIT Press).
—— (1990) 'A Theory of Content, II: The Theory', in *A Theory of Content and Other Essays* (Cambridge, MA: The MIT Press), 89–136.
R. Fogelin (1971) *Wittgenstein* (London: Routledge).
A. Ford (2008) *Action and Generality*, Ph. D. Dissertation, unpublished.
M. Forster (2004) *Wittgenstein on the Arbitrariness of Grammar* (Princeton, NJ: Princeton University Press).
P. Frascolla (2007) *Understanding Wittgenstein's Tractatus* (London and New York: Routledge).
G. Frege (1884) *Die Grundlagen der Arithmetik. Centenarausgabe* ed. Christian Thiel, Hamburg, Meiner 1986; English tr. by J.L. Austin, *The Foundations of Arithmetic*, 1953 (Oxford: Blackwell)
—— (1892) 'Über Sinn und Bedeutung', *Zeitschrift für Philosophie und philosophische Kritik*, 100, 25–50, English tr. in Frege (1967).
—— (1893) *Grundgesetze der Arithmetik*. Vol. 1 (Jena: Pohle).
—— (1918) *Logische Untersuchungen. Erster Teil: Der Gedanke*, in *Beiträge zur Philosophie des Deutschen Idealismus*. English tr. in Frege (1967).
—— (1967, 1990) *Kleine Schriften*, I. Angelelli (ed.) (Hildesheim: Olms). English edn. by B. McGuinness (1984) *Collected Papers on Mathematics, Logic, and Philosophy*, second edition (Oxford: Blackwell).
—— (1964) *The Basic Laws of Arithmetic: Exposition of the System*, tr. and ed. by M. Furth (Berkeley, CA: University of California Press).
—— (1969) *Nachgelassene Schriften*, H. Hermes et al. (eds.) (Hamburg: Meiner). English tr. by P. Long and R. White (1979) *Posthumous Writings* (Oxford: Blackwell).
—— (1980) *Begriffsschrift*, in P. Geach and M. Black (eds.), *Translations from the Philosophical Writings of Gottlob Frege*, 3rd edn. (Oxford: Blackwell), 1–20.
W.B. Gallie (1964) *Philosophy and the Historical Understanding* (London: Chatto and Windus).
M. García-Carpintero (2006), 'Two-Dimentionalism: A Neo-Fregean Interpretation', in M. García-Carpintero and J. Macià (eds.), *Two-Dimensional Semantics* (Oxford: Oxford University Press), 181–204.

P. Geach (1958) *Mental Acts* (London: Routledge).
—— (1972) *Logic Matters* (Oxford: Blackwell).
A. Gibbard (1994) 'Meaning and Normativity', in E. Villanueva (ed.), *Philosophical Issues 5: Truth and Rationality* (Atascadero, CA: Ridgeview), 95–115.
H.J. Glock (1996) *A Wittgenstein Dictionary* (Oxford: Blackwell).
—— (2000) 'Animals, Thoughts and Concepts', *Synthese*, vol. 123, 35–64.
—— (2003) *Quine and Davidson on Language, Thought and Reality* (Cambridge: Cambridge University Press).
—— (2004) 'Die Schreckliche Englische Sprache', in W. Lütterfelds und T. Mohrs (Hg.), *Die Welt ist meine Welt – Kulturelle Aspekte der Globalisierung* (Frankfurt: Peter Lang), 59–78.
—— (2005) 'The Normativity of Meaning Made Simple', in A. Beckermann and C. Nimtz (eds.), *Philosophy – Science – Scientific Philosophy* (Paderborn: Mentis), 232–3.
—— (2008) 'Meaning, Rules and Conventions', in E. Zamuner and D.K. Levy (eds.), *Wittgenstein's Enduring Arguments* (London: Routledge), 156–78.
S. Goldberg (2000) 'Do Anti-Individualistic Construals of Propositional Attitudes Capture the Agent's Conceptions?', *Noûs*, vol. 36, 597–621.
K. Glüer (1999) *Sprache und Regeln* (Berlin: Akademie Verlag).
N. Goodman (1951) *The Structure of Appearance* (Cambridge MA: Harvard University Press).
S.I. Greenspan and S.G. Shanker (2004) *The First Idea* (Cambridge MA: Da Capo Press).
P.M.S. Hacker (1972) *Insight and Illusion* (Oxford: Oxford University Press).
—— (1988) 'Language, Rules and Pseudo-Rules', *Language and Communication*, vol. 8, 159–72.
—— (1996a) 'Davidson on First-Person Authority', *Philosophical Quarterly*, vol. 47, 285–304.
—— (1996b) *Wittgenstein's Place in Twentieth-century Analytic Philosophy* (Oxford: Blackwell).
I. Hacking (1986) 'A Parody of Conversation', in E. LePore (ed.), *Truth and Interpretation* (Oxford: Blackwell), 447–58.
O. Hanfling (1989) *Wittgenstein's Later Philosophy* (London: Macmillan).
G. Harman (1999) *Reasoning, Meaning, and Mind* (Oxford: Clarendon Press).
H.L. Hart (1961) *The Concept of Law* (Oxford: Oxford University Press).
P. Horwich (1998a) *Truth*, 2nd edn (Oxford: Oxford University Press).
—— (1998b) *Meaning* (Oxford: Oxford University Press).
—— (2005) *Reflections on Meaning* (Oxford: Oxford University Press).
—— (2006) 'The Value of Truth', *Nous*, vol. 40, 347–60.
—— (2008) 'Ungrounded Reason', *Journal of Philosophy*, vol. 105, 453–71.
—— 'Kripke's Paradox of Meaning', forthcoming.
K. Hutchins, D.H. Hockenbury and S.E. Hockenbury (2002) *Telecourse Study Guide* to Accompany *Psychology* (US: Worth), http://coasttranscripts.org/pdfs/PHE_108.pdf
D. Kahneman and A. Tversky (1982) 'The Simulation Heuristic', in D. Kahneman, P. Slovic and A. Tversky (eds.), *Judgement under Uncertainty* (Cambridge: Cambridge University Press).
I. Kant, *Kritik der reinen Vernunft* (1787) English tr. by N. Kemp Smith (London: Palgrave Macmillan), 2003.

A.J.P. Kenny (1981) 'Wittgenstein's Early Philosophy of Mind', in I. Block (ed.), *Perspectives on the Philosophy of Wittgenstein* (Oxford: Blackwell), 140–7.
A.J.P. Kenny (1989) *The Metaphysics of Mind* (Oxford: Oxford University Press).
—— (1995) *Frege* (London: Penguin).
—— (2008) *From Empedocles to Wittgenstein* (Oxford: Oxford University Press).
J. Kim (1996) *Philosophy of Mind* (Boulder, CO: Westview Press).
—— (2003) 'Philosophy of Mind and Psychology', in K. Ludwig (ed.), *Donald Davidson* (Cambridge: Cambridge University Press), 113–36.
S.C. Kleene (1952) *Introduction to Metamathematics* (Amsterdam: North-Holland).
U. Kriegel (2007) 'Intentional Inexistence and Phenomenal Intentionality', *Philosophical Perspectives*, vol. 21, 307–40.
S. Kripke (1979) 'A Puzzle about Belief', in A. Margalit (ed.), *Meaning and Use* (Dordrecht: Reidel), 239–83.
—— (1980) *Naming and Necessity* (Oxford: Oxford University Press).
—— (1982) *Wittgenstein on Rules and Private Language* (Cambridge MA: Harvard University Press).
W. Ladusaw (1996) 'Negation and Polarity Items', in S. Lappin (ed.), *The Handbook of Contemporary Semantic Theory* (Oxford: Blackwell), 321–41.
G. Lando (2006) *Le relazioni nel Tractatus Logico-Philosophicus*, Ph.D. Thesis, unpublished.
M. Lance and J. Hawthorne (1997) *The Grammar of Meaning: Normativity and Semantic Content* (Cambridge: Cambridge University Press).
E. Lepore and K. Ludwig (2002) 'The Reality of Language', unpublished MS.
D. Lewis (1969) *Convention* (Oxford: Blackwell).
—— (1983) *Philosophical Papers*, vol. I (Oxford: Oxford University Press).
N. Malcolm (1958) *Ludwig Wittgenstein. A Memoir* (Oxford: Oxford University Press).
—— (1986) *Nothing is Hidden* (Oxford: Blackwell).
D. Marconi (1987) *L'eredità di Wittgenstein* (Rome: Laterza).
—— (1995) 'Fodor and Wittgenstein on Private Language', in R. Egidi (ed.), *Wittgenstein: Mind and Language* (Dordrecht: Kluwer), 107–15.
—— (1997) *Lexical Competence* (Cambridge MA: MIT Press).
—— (2005) 'Two-Dimensional Semantics and the Articulation Problem', *Synthese*, vol. 143, 321–49.
—— (2008) 'Conceivability', in R. Almeder (ed.), *Rescher Studies. A Collection of Essays on the Philosophical Work of Nicholas Rescher* (Frankfurt: Ontos Verlag), 169–95.
R. McDonough (1994) 'Wittgenstein's Reversal on the "Language of Thought" Doctrine', *The Philosophical Quarterly*, vol. 44, 482–94.
J. McDowell (1984) 'Wittgenstein on Following a Rule', *Synthese*, vol. 58, 325–63.
—— (1992) 'Meaning and Intentionality in Wittgenstein's Later Philosophy', *Midwest Studies in Philosophy*, vol. 17, 40–52.
M. McGinn (1997) *Wittgenstein and the Philosophical Investigations* (London: Routledge).
A. Mele (2003) 'Philosophy of Action', in K. Ludwig (ed.), *Donald Davidson* (Cambridge: Cambridge University Press), 64–84.
R. Moran (2004) 'Anscombe on Practical Knowledge', in J. Hyman and H. Steward (eds.), *Agency and Action* (Cambridge: Cambridge University Press), 43–68.
D. Moyal-Sharrock and W. Brenner (eds.) (2007) *Readings of Wittgenstein's On Certainty* (London: Palgrave Macmillan).

K. Mulligan and B. Smith (1986) 'A Relational Theory of the Act', *Topoi*, vol. 5, 115–30.
R. Nozick (2001) *Invariances: The Structure of the Objective World* (Cambridge MA: Harvard University Press).
C. Peacocke (1983) *Sense and Content: Experience, Thought, and their Relations* (Oxford: Oxford University Press).
—— (1992) *A Study of Concepts* (Cambridge MA: MIT Press).
—— (2004) *The Realm of Reason* (Oxford: Oxford University Press).
—— (2008) *Truly Understood* (Oxford: Oxford University Press).
D. Pears (1988) *The False Prison II* (Oxford: Clarendon Press).
E. Picardi (2002) 'Il principio del contesto in Frege e Wittgenstein', in C. Penco (ed.), *La svolta contestuale* (Milano: McGraw Hill), 1–23.
—— (2007) 'On Sense, Tone, and Accompanying Thoughts', in R.B. Auxier, L. E. Hahn (eds.), *The Philosophy of Michael Dummett* (LaSalle: Open Court), 491–520.
—— (2008) 'Frege and Davidson on Predication', in M.C. Amoretti, N. Vassallo (eds.), *Knowledge, Language and Interpretation. On the Philosophy of Donald Davidson* (Frankfurt: Ontos Verlag), 49–79.
—— (2009) 'Concepts and Primitive Language-Games', in E. Zamuner, D. Levy (eds.), *Wittgenstein's Enduring Argument* (London: Routledge), 109–34.
G. Priest, J.C. Beall and B. Armour-Garb (eds.) (2004) *The Law of NonContradiction: New Philosophical Essays* (Oxford: Clarendon Press).
A.N. Prior (1960) 'The Runabout Inference-Ticket', *Analysis*, vol. 21, 38–9.
H. Putnam (1975) 'The Meaning of "Meaning", in *Philosophical Papers II* (Cambridge: Cambridge University Press), 215–71.
—— (1988) *Representation and Reality* (Cambridge MA: The MIT Press).
—— (1990) 'Is Water Necessarily H_2O?', in J. Conant (ed.), *Realism with a Human Face* (Cambridge MA: Harvard University Press), 54–79.
W.V.O. Quine (1953) 'Two Dogmas of Empiricism' in *From a Logical Point of View* (Cambridge MA: Harvard University Press), 20–46.
—— (1960) *Word and Object* (Cambridge MA: Harvard University Press).
—— (1970) *Philosophy of Logic* (Englewood Cliffs, NJ: Prentice-Hall).
F. Recanati (2004) *Literal Meaning* (Cambridge: Cambridge University Press).
—— (2005) 'Literalism and Contextualism: Some Varieties', in G. Preyer and G. Peter (eds.), *Contextualism in Philosophy* (Oxford: Oxford University Press), 171–96.
E. Reck (1997) 'Frege's Influence on Wittgenstein: Reversing Metaphysics via the Context Principle', in W. Tait (ed.), *Early Analytic Philosophy. Frege, Russell, Wittgenstein* (Chicago and La Salle: Open Court), 123–86.
M. Richard (1997) 'Propositional Attitudes', in B. Hale and C. Wright (eds.), *A Companion to the Philosophy of Language* (Oxford: Blackwell), 197–226.
T. Ricketts (1996) 'Pictures, Logic, and the Limits of Sense in Wittgenstein's *Tractatus*', in H. Sluga and D.G. Stern (eds.), *The Cambridge Companion to Wittgenstein* (Cambridge: Cambridge University Press), 59–99.
G. Rizzolatti and M.A. Arbib (1998), 'Language within our Grasp', *Trends in Neuroscience*, vol. 21, 188–94.
R. Rorty (2009) 'Wittgenstein, Theory, and Therapy', in A. Ahmed (ed.), *Wittgenstein's Philosophical Investigations: A Critical Guide* (Cambridge: Cambridge University Press), forthcoming.

B. Rundle (1990) *Wittgenstein and Contemporary Philosophy of Language* (Oxford: Blackwell).
B. Russell (1912) *The Problems of Philosophy* (Oxford: Oxford University Press).
—— (1921) *Analysis of Mind* (London: Allen and Unwin).
—— (1940) *An Inquiry into Meaning and Truth* (London: Allen and Unwin).
—— (1956) *The Philosophy of Logical Atomism* in *Logic and Knowledge. Essays 1901–1950*, R.C. Marsh (ed.) (New York: The Macmillan), 175–282.
E. Sacchi and A. Voltolini (2009) 'To Think is to Have Something in One's Thought', unpublished MS.
M. Sainsbury (2005) *Reference Without Referents* (Oxford: Oxford University Press).
J. Saramago (2002) *The Cave* (New York: Harcourt).
S. Schroeder (1998) *Das Privatsprachenargument* (Paderborn: Schöningh).
J. Schulte (1982) 'Seguire una regola', *Lingua e Stile*, vol. 17, 497–512.
—— (1993) *Experience and Expression: Wittgenstein's Philosophy of Psychology* (Oxford: Clarendon Press).
—— (2007) 'Rules and Reason', *Ratio*, vol. 20, 464–80.
—— (2009) 'Concepts and Concept-Formation', Proceedings of a conference on 'Wittgenstein After His Nachlass' (Lisbon: New University), 2008, forthcoming.
J.R. Searle (1969) *Speech Acts* (Cambridge: Cambridge University Press).
—— (1983) *Intentionality* (Cambridge: Cambridge University Press).
—— (1992) *The Rediscovery of the Mind* (Cambridge, MA: The MIT Press).
N. Shah and D.Velleman (2005) 'Doxastic Deliberation', *Philosophical Review*, vol. 114, 497–534.
B. Smith (1998) 'Language, Conventionality of', in E. Craig (ed.), *Routledge Encyclopedia of Philosophy*, vol. 5 (London: Routledge), 368–71.
M. Smith (2004) 'The Structure of Orthonomy', in J. Hyman and H. Steward (eds.), *Agency and Action* (Cambridge: Cambridge University Press), 165–94.
S. Soames (1995) 'T-sentences', in W. Sinnott-Armstrong, D. Raffman and N. Asher (eds.), *Modality, Morality and Belief: Essays in Honor of Ruth Barcan Marcus* (Cambridge: Cambridge University Press).
—— (2002) *Beyond Rigidity* (Oxford: Oxford University Press).
—— (2003) *Philosophical Analysis in the Twentieth Century*, vol. 2, *The Age of Meaning* (Princeton, NJ: Princeton University Press).
—— (2005) 'Naming and Asserting', in Z.G. Szabò (ed.), *Semantics versus Pragmatics* (Oxford: Oxford University Press), 356–82.
E. Stenius (1960) *Wittgenstein's Tractatus. A Critical Exposition of Its Main Lines of Thought* (Oxford: Blackwell).
D.G. Stern (2004) *Wittgenstein's Philosophical Investigations. An Introduction* (Cambridge: Cambridge University Press).
F. Stoutland (1985) 'Davidson on Intentional Behavior', in E. Lepore and B. McLaughlin (eds.), *Actions and Events* (Oxford: Blackwell), 44–59.
—— (1999) 'Intentionalists and Davidson on Rational Explanation', in G. Meggle (ed.), *Actions, Norms and Values* (Berlin: Walter de Gruyter), 191–208.
—— (2009) 'Determinism, Intentional Action, and Bodily Movements', in C. Sandis (ed.), *New Essays on the Explanation of Action* (Basingstoke and New York: Palgrave Macmillan), 313–37.

—— (2010) 'Wittgenstein, Analytical Philosophy, and Metaphysics', unpublished MS.
P.F. Strawson (1959) *Individuals* (London: Methuen).
T. Szabó Gendler and J. Hawthorne (2002b), 'Introduction: Conceivability and Possibility', in T. Gendler Szabó, J. Hawthorne (eds.), *Conceivability and Possibility* (Oxford: Clarendon Press), 1–70.
A. Tarski (1983) 'The Concept of Truth in Formalized Languages', tr. J.H. Woodger, in J. Corcoran (ed.), *Logic, Semantics, Metamathematics*, 2nd edn. (Indianapolis, IL: Hackett), 152–278.
P. Tidman (1994) 'Conceivability as a Test for Possibility', *American Philosophical Quarterly*, vol. 31, 297–309.
M. Tomasello (1999) *The Cultural Origins of Human Cognition* (Cambridge, MA: Harvard University Press).
C. Travis (1989) *The Uses of Sense: Wittgenstein's Philosophy of Language* (Oxford: Oxford University Press).
—— (2006) *Thought's Footing. A Theme in Wittgenstein's Philosophical Investigations* (Oxford: Oxford University Press).
—— (2009) 'To Represent As So', in E. Zamuner, D. Levy (eds.), *Wittgenstein's Enduring Argument* (London: Routledge), 4–29.
A. Tversky and D. Kahneman (1983) 'Extension versus Intuitive Reasoning: The Conjunction Fallacy in Probability Judgment', *Psychological Review*, vol. 90, 293–315.
J. Van Cleve (1983) 'Conceivability and the Cartesian Argument for Dualism', *Pacific Philosophical Quarterly*, vol. 64, 35–45.
A. Voltolini (2005) 'On the Metaphysics of Internalism and Externalism', *Disputatio*, vol. 18, 127–50.
—— (2006) *How Ficta Follow Fiction. A Syncretistic Account of Fictional Entities* (Dordrecht: Springer).
E. von Savigny (1983) *Zum Begriff der Sprache* (Stuttgart: Reclam).
G.H. von Wright (1963) *Norm and Action* (London: Routledge and Kegan Paul).
—— (1971) *Explanation and Understanding* (Ithaca, New York: Cornell University Press).
—— (1982) *Wittgenstein* (Oxford: Blackwell).
K. Walton (1990) *Mimesis and Make-Believe* (Cambridge, MA: Harvard University Press).
F. Waismann (1976) *Logik, Sprache, Philosophie*, G.P. Baker and B.F. Mc Guinness (eds.) (Stuttgart: Reclam).
H. Wettstein (2004) *The Magic Prism* (Oxford: Oxford University Press).
D. Whiting (2007) 'The Normativity of Meaning Defended', *Analysis*, vol. 67, 133–40.
M. Williams (1988) 'Transcendence and Return: The Overcoming of Philosophy in Nietzsche and Wittgenstein', *International Philosophical Quarterly*, vol. 38, 403–19.
—— (1999) *Wittgenstein, Mind and Meaning: Toward a Social Conception of Mind* (London: Routledge).
—— (2004) 'Nonsense and Cosmic Exile: The Austere Reading of the *Tractatus*', in M. Kölbel and B. Weiss (eds.), *Wittgenstein's Lasting Significance* (London: Routledge), 1–27.

T. Williamson (1989) 'Review of *L'eredità di Wittgenstein*', *Philosophical Books*, vol. 30, 30–3.
T. Williamson (1998) 'Competenza lessicale e esternismo semantico', tr. F. Bertolucci, *Iride*, vol. 11, 397–401.
—— (2003) 'Understanding and Inference', *The Aristotelian Society*, suppl. vol. 77, 249–93.
—— (2006) 'Conceptual Truth', *The Aristotelian Society*, suppl. vol. 80, 1–41.
—— (2008) *The Philosophy of Philosophy* (Oxford: Blackwell).
—— (2009) 'Reference, Inference and the Semantics of Pejoratives', in J. Almog and P. Leonardi (eds.), *Essays for David Kaplan* (Oxford: Oxford University Press), forthcoming.
C. Wright (1989) 'Wittgenstein's Rule-Following Considerations and the Central Project of Linguistics', in A. George (ed.), *Reflections on Chomsky* (Oxford and New York: Blackwell), 233–64, later in Wright 2001.
—— (2001) *Rails to Infinity* (Cambridge MA: Harvard University Press).
—— (2002) 'The Conceivability of Naturalism', in T.Szabó Gendler, J. Hawthorne (eds.), *Conceivability and Possibility* (Oxford: Clarendon Press), 401–39.
—— 'What is Wittgenstein's Point in the Rule-Following Discussion?', unpublished MS.
S. Yablo (1993) 'Is Conceivability a Guide to Possibility?', *Philosophy and Phenomenological Research*, vol. 53, 1–42.
—— (2002) 'Coulda, Woulda, Shoulda', in T. Szabó Gendler, J. Hawthorne (eds.), *Conceivability and Possibility* (Oxford: Clarendon Press), 441–92.

Index